Marine Pollution

Marine Pollution

FOURTH EDITION

R. B. Clark

Emeritus Professor of Zoology
University of Newcastle upon Tyne

in collaboration with

Chris Frid

Dove Marine Laboratory, University of Newcastle upon Tyne

and

Martin Attrill

Department of Biological Sciences, University of Plymouth

CLARENDON PRESS · OXFORD

Oxford University Press, Great Clarendon Street, Oxford OX2 6DP

Oxford   New York

Athens   Auckland   Bangkok   Bogota   Bombay   Buenos Aires   Calcutta
Cape Town   Chennai   Dar es Salaam   Delhi   Florence   Hong Kong   Istanbul
Karachi   Kuala Lumpur   Madrid   Melbourne   Mexico City   Mumbai
Nairobi   Paris   São Paulo   Singapore   Taipei   Tokyo   Toronto   Warsaw

and associated companies in
Berlin   Ibadan

Oxford is a trade mark of Oxford University Press

Published in the United States
by Oxford University Press Inc., New York

© R. B. Clark, 1986, 1989, 1992, 1997

First published 1986
Second edition 1989
Third edition 1992
Fourth edition 1997, Reprinted 1998

A catalogue record for this book is available from the British Library

Library of Congress Cataloging in Publication Data
(Data available)

ISBN 0 19 850070 X (Hbk)
ISBN 0 19 850069 6 (Pbk)

Printed in Great Britain by
The Bath Press, Bath

PREFACE

The objective of university teaching is partly to give students access to sources of current information, but also, importantly, to encourage them to think independently and form judgements which they can justify and support with tested evidence. Environmental pollution is a subject that can present special difficulties in this respect because students often have preconceptions reflecting current popular wisdom. The current popular assessment of the significance of pollution in the sea may be right or wrong, but it and the grounds on which it is based must not be above dispassionate examination. A few years ago, as noted in the preface to the third edition of this book, many students were committed environmentalists already holding strong views. To them, dispassionate examination of the facts was either irrelevant, or worse, indicated an 'anti-environmental' attitude. In those circumstances, the teacher, whether subscribing to the conventional wisdom or not, was in danger of being accused of trying to indoctrinate the student, and that was not a happy situation when the objective of the instruction was to encourage rational and objective thought. This is less of a problem today. Younger students, in particular, are often unaware of the problems associated with waste disposal to the sea, or imagine that governments now have these problems in hand, but the need for a dispassionate review of the subject is no less necessary.

The scientific appraisal of the health of the seas made a decade ago, when the first edition of this book was published, has proved surprisingly reliable. The most notable change during that time has been in public opinion, not in science. In the early 1980s, there were voices proclaiming that the sea was being poisoned by pollution; indeed, some claimed that the marine environment had already been irreparably damaged by our waste discharges. To a considerable extent, these pessimistic voices were disregarded and dumping of wastes into the sea continued on a large scale in most parts of the world. Now, environmental expectations are high and continue to rise, and waste disposal practices have had to be adjusted to take account of this change of public attitude. The marine environment is robust and recovery of damaged areas has followed these changed practices. This does not prevent alarmist cries of 'catastrophe' and 'permanent damage' each time there is a spectacular accident, nor reduce the strong public pressure, often irresistible, that 'something be done', even though remedial action may be useless, or worse, increase the damage. In environmental matters, emotion still too often prevails over reason.

Scientific perspectives of marine pollution have changed dramatically as progress has been made in reducing or removing old problems, and new problems have emerged. The new edition of this book takes account of these changes of perspective and fully reflects current concerns which, in many respects, are different from those discussed in previous editions. For example, ten years ago, eutrophication was not regarded as a problem in the sea, but is now regarded as the over-riding threat to the health of inshore waters in several parts of the world. Accordingly, it has been given more extensive treatment than in previous editions of this book. Previous editions tended to focus particularly on European waters since the problems there were generally better known than those in other seas. More information is now available, and it has been possible to draw more case-histories from other parts of the world as well as pay more attention to pollution problems in tropical seas. As in other revisions of the text, the opportunity has been taken to update factual information to include the most recently available data.

Parts of the book have been substantially changed to meet the needs of students. Recognizing that it is now widely used as a text to accompany lecture courses and is unlikely to be read consecutively, a more formal structure of the chapters has been adopted. In particular, a chapter on methodology has been introduced incorporating the discussion of toxicity testing which was previously treated in the chapter introducing conservative pollutants, and also introduces statistical treatments of field data, replacing some of the matters previously discussed in the final chapter. The introductory chapter has been expanded to include a discussion of approaches to pollution control.

The core chapters of the book retain the original approach of giving an objective, factual account of the wastes that enter the sea, what happens to them there, the effects they have on plants and animals living in the sea, and the consequences for human health. Judgements are not offered and readers are left to draw their own conclusions. Older case-histories which illustrate this account have been replaced where possible by more recent ones, and pollution events that have attracted popular attention in recent years are discussed, such as the *Exxon Valdez* oil spill in Alaska and the wreck of the tanker *Braer* on the south coast of Shetland, the disposal at sea of redundant offshore oil platforms like the *Brent Spar*, and Russian practices of radioactive waste disposal at sea, including the dumping of redundant nuclear submarines. More information is given than in previous editions about the effects of dredging, the disposal of solid wastes and the construction of artificial islands. Increased attention that has been paid to pollution abatement and marine conservation in most parts of the world, and the chapter on The State of Some Seas has been revised to take account of our greatly improved knowledge of these areas.

It has been suggested that students require better guidance to the scientific literature than that provided in the brief reading list in previous editions. It would clearly be inappropriate to lard a text-book with references to citations in an extended bibliography, but the section on Further Reading has been expanded to include a handful of key references relating to each chapter, and attention is drawn to the sources of all text-figures and tables which are now listed in that section. These will lead the reader to detailed information about all the case histories included in the text.

In preparing this edition, I have enjoyed the collaboration of Dr Chris Frid and Dr Martin Attrill. Both have extensive and current experience of teaching courses related to marine biology and marine pollution, and are attuned to the needs of students. Their experience has been crucial in designing this edition to meet those needs, but beyond that, they have made important contributions to the technical treatment of several sections of the text.

Newcastle upon Tyne   R. B. Clark
January 1997

ACKNOWLEDGEMENTS

The authors are most grateful to the following for advice, guidance and information on a variety of matters. Dr Jenifer Baker, Professor Barbara E. Brown, Dr John Bythell, Mr W. C. Camplin, Dr K. L. Clarke, Dr J. Coggins, Mr Simon Cullins, Dr Mike Elliott, Dr Lars Føyn, Dr Peter Gough, Miss Catherine Gray, Dr John Hall, Dr Richard Johnson, Dr Paul Johnston, Dr Paul Kingston, Miss Christina Lye, Mr David Moulder, Dr John B. Pearce, Dr David M. Porter, Dr Ian White and Dr J. Widdows.

CONTENTS

WHAT IS POLLUTION?

SOME QUESTIONS

Everyone knows what pollution is and that it is a 'bad thing', but for a scientific examination of marine pollution, or any other sort of pollution, value judgements of this kind have to be quantified. In what way is it bad? How bad? Bad for whom?

To answer these questions, we must consider:

- what kind of materials are discharged into seas or estuaries, or otherwise get there as a result of human activities;

- what effect these additions to the sea have on the marine or estuarine environment and the plants and animals living there;

- what implications these effects have for human health, food resources, commercial interests, amenities, wildlife conservation, or ecosystems in general;

- what is being done, can be done, or should be done to reduce or remove the damaging or undesirable effects of these additions to the marine environment;

- what would be the consequences of not releasing these materials to the sea and would such consequences be better or worse than the existing situation.

A quantitative, step-by-step approach to this subject is clearly required, and the rest of this book is concerned with answering these questions in quantitative terms and assessing the importance of various kinds of pollution impact. Before embarking on that, it is neces-sary to look at what types of materials enter the sea as a result of human activities.

CATEGORIES OF ADDITIONS

Degradable wastes

By far the greatest volume of wastes entering coastal waters and estuaries is composed of organic material which is subject to bacterial attack. Essentially, this is an oxidative process and ultimately breaks down organic compounds to stable inorganic compounds such as CO_2 (carbon dioxide), H_2O (water), and NH_3 (ammonia).

Wastes included under this heading are:

- a large part of urban sewage;

- agricultural wastes (now in substantial quantities where factory farming is practised);

- food processing wastes from slaughter houses and freezer plants, pulp from sugar beet factories, and so on;

- brewing and distillery wastes;

- paper pulp mill wastes which include much wood fibre;

- chemical industry wastes, including a great variety of large molecules which are relatively unstable and readily broken down;

- oil spillages.

In principle, such degradable wastes are no different from plant and animal remains which are subject to bacterial decay. Since bacteria are an important base of many food chains in the sea, the addition of organic matter represents an enrichment of the ecosystem, comparable to adding stable manure as a fertilizer in the garden.

If the rate of input exceeds the rate of bacterial degradation, organic materials **accumulate**. The rate of bacterial action depends on temperature, oxygen availability, and other factors; if these become limiting, the rate of bacterial action falls and the capacity of the waters to receive organic wastes without accumulation is much reduced. If the input of wastes is large, there is intense bacterial activity until the oxidative processes of degradation outrun the supply of oxygen dissolved in the water, leading to **deoxygenation**. In these circumstances, further degradation depends on the activity of **anaerobic** bacteria, which is slow and yields end-products such as hydrogen sulphide and methane.

Accumulation of organic materials and deoxygenation of the water both have a strong impact on the flora and fauna, and at very low oxygen levels most plants and animals are excluded. Thus, if the input of organic wastes is within the capacity of the receiving waters —which is related to temperature, oxygen availability, water currents, and so on—it will result in enrichment, of benefit chiefly to plants in the first instance. If the capacity of the receiving waters is exceeded, the accumulation of organic material and the development of anoxic conditions result in impoverishment of the fauna and flora.

Fertilizers

Agricultural fertilizers may have a similar effect to organic wastes. Nitrates and phosphates are leached from arable land and carried by rivers to the sea, where the fertilizers enhance phytoplankton production, sometimes to the extent that the accumulation of dead plant remains on the seabed produces anoxic conditions.

Dissipating wastes

A number of industrial discharges into the sea and estuaries rapidly lose their damaging properties after they enter the water. Any effects they may have are therefore confined to the area immediately around the point of discharge, although the extent of that area depends on the rate of discharge, water currents, and so on.

- *Heat*. This comes principally from the cooling water from coastal power stations and factories, although some other industrial effluents may be heated. Commonly, the discharge is at about 10 °C above the temperature of the receiving water. Dissipation of the heat depends chiefly on mixing of the hot water with cold water. In temperate seas heated discharges are generally of little consequence, but in tropical seas, where summer temperatures are already near to the thermal death point of many organisms, the increase in temperature can cause substantial loss of life.

- *Acids and alkalis*. Seawater has a large buffering capacity and the effect of such discharges is extremely localized.

- *Cyanide*. This comes principally from metallurgical industries. Cyanide rapidly dissociates in seawater and has little effect except in the immediate neighbourhood of the outfall.

Conservative wastes

Some materials are not subject to bacterial attack and are not dissipated, but are reactive in various ways with plants and animals, sometimes with harmful effects. Because of their persistence and harmful effects, they are regarded as a very serious threat. The principal categories of such wastes are:

- heavy metals (mercury, copper, lead, zinc and so on);

- halogenated hydrocarbons (DDT and other chlorinated hydrocarbon pesticides, polychlorinated biphenyls (PCBs) and so on);

- radioactivity.

Solid wastes

Inert solid wastes are a growing problem. These include litter, much of it composed of man-made plastics, including polythene containers, plastic sheeting, nylon ropes, nets, and other fishing gear. Polystyrene spherules, which are feedstock for the plastics industry, also reach the sea through accidental spillages and are now widespread in the world's oceans.

Finer particulate matter suspended in the water may clog the feeding and respiratory structures of animals, reduce plant photosynthesis by reducing light penetration, and, when it settles on the seabed, smother animals and change the nature of the seabed. Such materials include:

- dredging spoil;
- mining waste from dumping or coastal dredging for minerals;
- other industrial waste such as powdered ash (fly ash) from power stations and china clay waste;
- clay from gravel extraction; gravel dredged from the seabed contains a proportion of clay and silt which is washed out in the course of extraction and settles back on the seabed.

NATURE OF INPUTS

Although it is convenient to consider materials added to the sea as being in the above categories because of their shared properties and the similarity of the effects they have, it must be remembered that actual inputs to the sea are rarely so simple in their constitution.

- Power station effluent is chiefly hot water, but also contains some chlorine injected into the water at the intake to discourage marine organisms settling in the cooling system and reducing its efficiency or blocking it. Small amounts of metals are leached from the cooling system and turbines.
- Urban sewage is principally organic but also contains considerable amounts of metals, oils and greases, detergents, and industrial wastes (since most industrial plants use the same sewers as domestic users), as well as pathogens.

Similar complexity may exist in particular geographical sites. An industrialized estuary has a multiplicity of inputs from surrounding industry, often in great variety, as well as from the urban population. In addition, the estuary receives whatever is carried by the river flowing into it, which may include pesticides and other products of agricultural activity over the area of the entire catchment system.

SOURCES OF INPUTS

Direct outfalls

The most obvious inputs of material to the sea are through pipes discharging directly into it.

- *Estuaries.* Historically, most ports grew up on estuaries and became centres of population and industry. The urban and industrial wastes were discharged directly into the estuary without treatment. With the growth of population and industry, the major industrialized estuaries (Thames, Mersey, Meuse, Scheldt) became foul, stinking, and lifeless by the end of the nineteenth century.
- *Coastal towns.* Many coastal towns have little industry but, in the past, untreated municipal waste from them was discharged directly into the sea by numerous outfalls, sometimes at the high water mark and rarely far below low water. This resulted in fouling of local beaches, and many resorts where tourism is important have extended the outfalls farther offshore to reduce fouling, or have introduced sewage treatment.
- *Coastal industry.* With increasing pressure to conserve rivers and lakes for drinking water supplies, new industry, with demands for large volumes of water (for example, for cooling water and waste disposal), have tended to be sited on the coast. Coastal mariculture installations, such as salmon farms, are responsible for a considerable direct input of unconsumed food, fish excreta, and pesticides to inshore waters.

River inputs

Rivers flow via their estuaries to the sea and

potential pollutants reach rivers over the entire catchment area.

- Organic wastes are subject to bacterial attack in transit, and, depending on the distance of major inputs from the river mouth, the organic load entering the sea from upstream is correspondingly reduced and may be negligible.
- Pesticides and fertilizers from agriculture and forestry are washed off the land by rain, enter water courses, and eventually reach the sea.
- Petroleum and oils washed from roads by rain enter the sewerage system as a rule, but storm-water overflows carry these materials into rivers and to the sea.

All land-based materials washed off by rainfall and entering rivers and streams contribute inputs to the sea, and, in aggregate, they may form a very large contribution.

Shipping

Ships carry many toxic substances in the course of trade: oil, liquefied natural gas, pesticides, industrial chemicals, and so on. Shipwrecks and other accidents at sea may release these substances, and the very large size of some vessels—crude oil carriers of up to 350 000 tonnes, for instance—means that when accidents do occur, the consequences may be very damaging. Major oil tanker wrecks are an obvious example.

Other shipping is responsible for less dramatic inputs of materials to the sea, both as a result of accidents and in the course of routine operations. A ship does not have to be wrecked and become a total loss for this to happen; in all but the worst accidents much of the cargo is salvaged, but particularly noxious or dangerous materials are frequently carried as deck cargo as a safety precaution for the ship, and these may be lost overboard in severe storms.

In the course of routine operations, ships discharge oily ballast water and bilge water (not always legally) and cargo tank washings; they also discard much litter, of which plastics probably constitute the worst nuisance.

Offshore inputs

A variety of material is dumped at sea in designated dumping grounds.

- *Dredging spoil.* Shipping channels in estuaries and at the entrances to ports may require frequent dredging to keep them open, and the dredged material is barged out to sea and dumped. This dredging spoil, particularly from industrialized estuaries, may contain appreciable quantities of heavy metals and other contaminants, which are then transferred to the dumping grounds.

- *Sewage sludge.* Sludge from sewage treatment plants was formerly dumped at sea in considerable quantities, but the practice has now ceased in many areas. The sludge has a high organic content and is also contaminated with heavy metals, oils and greases, and many other substances.

- *Other wastes.* Fly ash from oil- and coal-fired power stations, colliery waste, and other industrial particulate wastes have been dumped at sea in a number of places; as have various dangerous materials such as radioactive waste and unwanted munitions.

- *Offshore industrial activities.* These result in a variety of inputs or disturbances in the sea. They include offshore oil exploration and extraction, sand and gravel extraction, and mining for minerals, including, in the future, manganese nodule extraction.

Atmospheric inputs

Discharges to the atmosphere are returned to the land or the sea in rain, or, if particulate in nature, as fallout. Gaseous wastes dissolve directly in the sea at its surface. Such inputs are, of course, on a regional or even global scale rather than local. There are too many unknown factors for atmospheric inputs to the sea to be estimated precisely, but they are generally supposed to be very large and to represent major contributions.

The total global input of lead to the sea from natural and man-made sources is estimated to be about 400 000 t year^{-1}. Of this, about half is derived from vehicle exhausts containing leaded petrol additives which reach the atmosphere and are then rained out.

The total global input of mercury to the sea from volcanic activity and weathering of mercury-bearing ores is at least 5000 t year^{-1}, but may be as high as $25\,000$ t. A further 5000 t year^{-1} are added from the use of mercury and mercurial compounds in industry and agriculture. An additional 3000 t are derived from burning fossil fuels, principally coal. Although coal contains only minute amounts of mercury, it is burned in such quantity that, in total, it makes a major contribution of mercury to the sea.

Such global figures are estimates only and cannot be regarded as accurate measurements, but it is clear that materials discharged to the atmosphere can make unexpectedly large contributions to the budget of these contaminants in the sea.

These two examples also reveal that there is often a substantial natural input of materials which, in some circumstances, prove to be pollutants.

DEFINING POLLUTION

The word 'pollution' is commonly used to mean:

- the environmental damage caused by wastes discharged into the sea;
- the occurrence of wastes in the sea;
- the wastes themselves.

This is confusing and does not encourage a detailed analysis of the effects of the wastes in the sea. Most pollution scientists use different terms for the wastes ('inputs'), the occurrence of them in the sea ('contamination'), and the damaging effects they have ('pollution'). These distinctions are also recommended by such international advisory bodies as the United Nations Group of Experts on the Scientific Aspects of Marine Pollution (GESAMP) and the International Commission for the Exploration of the Sea (ICES).

Inputs

Some of the wastes reaching the sea (for example, many pesticides, plastics) are man-made and do not otherwise occur in nature, but most of the substances discussed so far exist naturally in the sea:

- organic material subject to bacterial degradation;
- metals in the runoff from metalliferous deposits;
- oil from natural seeps such as those occurring in parts of the Gulf of Mexico, on the Californian coast, and parts of the British coast;
- particulate material from coastal erosion— on the north-east coast of England this includes coal from seams exposed on coastal cliffs;
- hot water from geothermal springs around the coast of Iceland, in the Galapagos Trench, and elsewhere;
- radioactivity on the coasts of Brazil and south-west India.

Because of this variety of natural sources, it is confusing to refer to inputs to the sea as 'pollution'. Are all inputs, natural as well as those resulting from human activities, to be regarded as polluting? Or are only human inputs polluting, even though natural inputs of the same substances may be very much greater?

Contamination

With so many natural inputs to the sea in different parts of the world, the concentrations of substances vary widely from place to place in the marine environment. **Contamination** is caused when an input from human activities increases the concentration of a substance in seawater, sediments, or organisms above the natural background level for that area and for the organisms. This locally elevated concentration may, of course, be less than the concentration of the same substances in other areas where there is a large natural input.

The sensitivity of modern chemical analytical techniques now makes it possible to detect extremely low concentrations of many substances in the sea. The apparent spread of contamination in recent years is due more to the improvement in analytical techniques than to any change in the quantity of materials entering the sea.

Many substances change their nature when added to seawater: compounds may dissociate or become ionized; metals may change their valency or form complexes with organic molecules; substances may dissolve in seawater or become adsorbed on to particulate matter and be carried to the seabed. These physico-chemical changes affect the degree to which the added substances are available to marine organisms and the effects they have on them.

For these reasons, a simple measure of the concentration of a substance in the sea is unlikely to reflect the effects it may have.

Pollution

The matter of greatest concern is, of course, the effects the inputs have in the marine environment. GESAMP defined pollution to reflect this aspect of wastes in the sea. **Marine pollution** is the introduction by man, directly or indirectly, of substances or energy to the marine environment resulting in deleterious effects such as: hazards to human health; hindrance of marine activities, including fishing; impairment of the quality for the use of seawater; and reduction of amenities. The focus is therefore on human rather than natural inputs to the sea, and on the damaging effects of wastes.

Practical implications

It can be argued that, with the advance of scientific knowledge, what at one time was believed to be a harmless level of contamination (that *is not* 'polluting') may later be found to cause subtle damage (it *is* 'polluting'). As a precaution against such unpleasant surprises, it would be prudent not to discharge any wastes into the sea.

Even if this policy is adopted, it does not remove the need to distinguish between inputs, contamination, and pollution. Many inputs to the sea are not deliberate but are derived from atmospheric fallout, runoff from the land, or accidental spillages. In addition, although the discovery of high concentrations of a substance may provide a warning signal, unless it is the result of human activities and is damaging, it does not constitute pollution. It should be noted, too, that even if laboratory studies show that a contaminant is toxic to particular marine organisms, this may provide another—possibly stronger—warning signal, but does not necessarily prove that the contaminant has harmful effects in the natural environment.

From a strictly biological point of view, even if toxic contaminants do cause the death of some plants and animals in the natural environment, this is usually of less consequence than if the deaths result in a change in the population as a whole. Most marine animals reproduce on a colossal scale and the overwhelming majority of offspring die prematurely in the natural course of events. Mortality from toxic contamination may be insignificant compared with these natural losses and may have no effect on the population.

It follows from these considerations that, whereas there can be no doubt about the existence of pollution when the damage is severe, there is often considerable difficulty in identifying damage from low levels of contamination caused by diffuse, or sometimes unknown, inputs. Pollution is by no means as clear-cut as might at first appear.

PRIORITIES

It will be noticed that in defining pollution, most examples of its deleterious effects relate to human interests in the sea. Pollution is, in practical terms, an example of one set of human interests:

- the use of leaded petrol in cars;
- burning coal;
- transporting oil;
- generating electricity;
- disposing of waste products

coming into conflict with other human interests:

- human health;
- amenities, tourism, recreation, and aesthetic values;
- scientific interest.

Generally, we would consider transport, the provision of electricity, and many vital industrial processes to be as important as preserving a fish supply or providing tourist facilities. So, we are really concerned with striking the most favourable balance (for humans) between these conflicting interests—in other words, we are concerned with priorities.

As a rule, a threat to human health is given the highest priority for abatement or control. There are too many examples to the contrary to claim that it is given priority over all other considerations, but it is certainly an issue that is considered to be of greatest importance.

When only two interests conflict, and both are commercial, it is fairly easy to decide the priorities. When the oil tanker *Torrey Canyon* was wrecked off the Cornish coast in 1967 and deposited a large amount of oil on the beaches, a decision had to be taken whether or not to treat the spilled oil with chemical dispersants, which, at that time, were very toxic to marine organisms: their use would have meant a serious risk of great damage to the local fishing industry, but to leave the oil untreated would have been to the great detriment of the tourist industry. The fishing industry was then worth £6 million per year and the tourist industry £60 million per year, so on that basis the decision was taken to clean the beaches. Happily, in the event, fears about damage to the fishing industry proved to be unfounded and it was not noticeably affected.

Generally, however, the issues are not so clear-cut and the decisions are more difficult. A more sophisticated approach is needed and, for this, a great deal of information is required.

- What is the level of contamination in the area we are interested in?
- What form does it take?
- Where does it come from?
- What happens to it in the sea?

- What does it do to the plants and animals there?
- If plants and animals are affected, does it matter?
- To whom does it matter? What other interest is affected?
- How much does it matter to them?
- If it does matter, what can we do about it?
- What do we do with the polluting material if it is not put in the sea?
- Would the alternative be better or worse than putting it in the sea?
- How much would it cost?

COST OF POLLUTION ABATEMENT

A number of these questions involve matters that are economic or political and are outside the realm of pure science. Waste disposal, however it is managed, has both an environmental and a financial cost. While we wish to minimize the environmental cost, financial costs cannot be ignored because, depending on the environmental standards demanded, they can escalate very rapidly, as the following examples show.

The capital cost of building a new coal-fired power station equipped to remove particulates from the flue gas is as follows:

- 90 per cent extraction of particulates adds 10 per cent to the cost;
- 95 per cent extraction adds 20–30 per cent to the cost;
- 99 per cent extraction doubles the capital cost of the power station.

A sugar beet processing plant handling 2700 t day^{-1} produces an effluent high in organic wastes. The organic content is measured as BOD (biochemical oxygen demand, see p. 23) and this can be reduced by suitable treatment, but at a cost:

- 30 per cent reduction in BOD costs less than $0.50 kg^{-1};

- 65 per cent reduction costs $10 kg^{-1};
- 95 per cent reduction costs $30 kg^{-1}.

A similar exponential relationship between effluent standards and treatment costs exists for most wastes (Fig. 1.1); at the margin, costs can easily double for a trivial improvement in environmental quality. A decision has to be made as to where on the exponential curve a limit should be set.

Scientists can probably predict fairly accurately the effect on the local environment of a particular discharge, but they cannot decide on an appropriate trade-off between environmental cost and financial cost. That must be a political decision, which is made by taking into account other matters such as the value placed on the local environment, public perceptions of risks, and social costs if a factory is forced out of business by high effluent treatment costs.

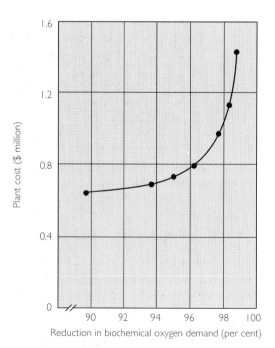

Fig. 1.1 Capital cost of treating an effluent containing organic and industrial waste. (*Published with the permission of the Controller of Her Majesty's Stationery Office.*)

APPROACHES TO POLLUTION CONTROL

Pollution control measures are certainly not always undertaken on a rational basis. Remedial action is sometimes instituted before it is known whether a waste product is damaging in any material way, or if the action taken will have a beneficial effect. Aroused, but ill-informed public opinion may insist that 'something' be done to remedy a situation, and an industry or government body may be obliged to undertake completely unless activities so that they can be seen to be active. Such activities may be harmless, but they are expensive and wasteful of resources that might be better spent for some other purpose.

In other respects, irrational attempts at pollution control may be positively harmful. Awareness of possible dangers stemming from pollution of the sea has encouraged the view that nothing should be discharged or dumped into it. But this is not a solution, it merely transfers the problem of waste disposal elsewhere. Care has to be taken to ensure that in solving one problem we are not simply creating a worse one in a different environment.

If waste disposal is to be carried out rationally and with due concern for the environment, it is necessary to establish principles for the disposal of wastes that result in the least environmental damage. A number of such approaches underlie attempts to bring pollution in the sea under control.

Banning pollutants

When a substance is clearly identified as being unacceptably damaging, it may be possible to withdraw it from use altogether. A number of products have been excluded in this way, but this approach to pollution control has severe limitations.

- Care has to be taken that the replacement for the banned product is not itself damaging to the environment. Withdrawing DDT, mercurial wood preservatives, or copper-based antifouling paints, as well as other substances, have all presented unforeseen problems.

• Withdrawing a product from use does not always provide a remedy to environmental problems in the short term if, like persistent pesticides and PCBs, discharges that have accumulated in the environment for several decades continue to cause damage.

• It is impracticable to withdraw some environmentally damaging substances from use. The distressing loss of seabirds from oil pollution can be stopped only by ceasing to use and transport oil at sea. That would eliminate the use of oil and oil products in many countries and effectively ground all aircraft; that would be regarded as too high a price to pay.

Best practicable environmental option

Industrial societies inevitably produce a great variety of waste in large quantities. By fostering industrial processes that produce little or no waste (**low waste** and **no waste technology**) and by withdrawing particularly damaging substances from use, it is possible to reduce pollution. Even so, large quantities of waste will always be produced (for example, sewage) and have to be disposed of somewhere; if not in the sea then in another environment. Waste disposal always has some impact on the environment, and in dealing with the pollution problems in one environment—the sea—it is only too easy to suggest solutions that merely transfer the problem to a different environment.

In countries like Britain or The Netherlands, which are densely populated and where land is relatively scarce, it is important not to solve a problem in the sea by creating a greater one on land. What is required is the means of disposing of a waste that will cause the least environmental damage. This is known as the **Best Practicable Environmental Option**. It may mean that, in some instances, disposal of a waste to sea, even though it causes some damage to marine resources, is preferable to any other method of disposal.

Environmental impact assessment

In most developed countries, waste discharge requires some kind of official consent or licence. When a new discharge is proposed, the discharger must provide a detailed **Environmental Impact Assessment** (EIA) or **Environmental Impact Statement** (EIS) which predicts the effect of the proposed discharge on the surrounding area. On the basis of this, a permit may be issued or denied, or modifications of the discharge be required. The requirement to provide EIAs has been in force in North America for a number of years. In countries where a formal assessment is not required, very much the same inquiry is made before a permit to discharge a waste to sea is granted.

A recent analysis of the accuracy of the predictions made in EIAs submitted in Australia has shown that slightly more than fifty per cent of the assessments underestimated or overestimated the environmental impact of the developments. There have been no other surveys of the reliability of EIA, but it is unlikely that the experience in Australia is unique.

Environmental capacity

While the best practicable environmental option approach to waste disposal accepts that some damage may be acceptable in the marine environment, the concept of environmental or assimilative capacity provides a formula for limiting that damage.

At some level, the marine environment obviously has the capacity to tolerate an input of man-made wastes without suffering significant damage. That level may be quite small for some wastes, and different sea areas will have different capacities to dilute and disperse wastes to a harmless level. In addition, there will obviously be differences of opinion about what is significant or acceptable damage. Nevertheless, in principle, it is possible to prescribe an appropriate level of waste discharge to a sea area that is within its capacity to assimilate.

Assimilative capacity is a well-known concept in freshwater and underlies the calculation of safe discharges of sewage wastes to rivers. The existence of water of a quality suitable for fish, or for abstraction as a drinking water supply downstream of the input, is clear evidence of the success of this approach.

The marine environment is more compli-cated than a river system and our understand-ing of processes in the sea is far from complete, so it is difficult to estimate its assim-ilative capacity with certainty. For this reason, when a waste discharge or other undertaking that is likely to cause some disturbance in a sea area is planned, a regulatory system involving some kind of 'feedback loop' is desirable. By this means, the response of the ecosystem is measured to ensure that it is within the predicted (and acceptable) con-sequences of the input, or, if not, to modify whatever is causing the disturbance. This kind of monitoring commonly accompanies major construction works on the coast or in the sea

Best available technology

Another approach to reducing environmental damage in the sea is to insist that waste dis-charges should be treated by the **best avail-able technology** (BAT) to minimize the release of damaging substances. This may seem a logical principle, but it has proved to be controversial.

The use of the best available technology may be over-protective if a small quantity of non-persistent waste is discharged into a large body of water, or it may give inadequate pro-tection if several different sources discharge into a limited environment; the technology used to remove the noxious constituents of the waste may result in damaging con-sequences in a different environment; and, finally, strict insistence on the use of the best available technology pays no regard to the cost of the treatment.

The high, and sometimes unnecessary, cost of using the best available technology has led

to some resistance to its widespread introduc-tion. As a compromise, the **Best Available Technology Not Entailing Excessive Costs** (BATNEEC) or the **Best Practicable Tech-nology** (BPT) have been widely accepted, although they offer weaker protection of the marine environment.

Precautionary principle

Fears about the damaging effects of wastes entering the North Sea led to the develop-ment of the **precautionary principle**, which was proposed by Germany in 1986. It argues that, since we cannot reliably predict the effect of new inputs added to an area that already receives a large volume of wastes, no wastes should be discharged to the sea unless they can be shown to be harmless. This is a very stringent requirement because it is almost impossible to prove that a waste is harmless and, indeed, there are very few wastes that will not cause some environmental adjustment where they are discharged.

While the principle has received general acceptance, at least in a somewhat less strin-gent form, it does not address the fact that river and atmospheric inputs to the North Sea are often more significant than direct dis-charges. Furthermore, it is concerned solely with protecting the sea and does not take into account the environmental consequences of disposing of the wastes in a different environ-ment. One major criticism of the way in which the principle has sometimes been applied is that the slightest suspicion that a waste had a damaging effect was accepted as sufficient to warrant its control, without any consideration of scientific evidence. If the principle was strictly applied, there would be no place for science in decision making.

MEASURING CHANGE

Any material discharged into the sea inevitably causes some change in the environment. The change may be great or small, long lasting or transient, widespread or extremely localized. If the change can be detected and is regarded as damaging, it constitutes pollution. Much effort is devoted to measuring levels of contamination of sediments and organisms by chemical analysis, but to determine if the observed level of contamination causes pollution generally requires a study of its biological effects. These effects may be detected at the level of the individual, or by changes in the population or the community, and a variety of techniques is available to identify and measure the responses.

IMPACT ON THE INDIVIDUAL

Individual organisms may suffer some form of impairment or damage as a result of exposure to a pollutant. In extreme cases it results in death.

Measurement of toxicity

Toxicity states how poisonous a substance is, or how large a dose is required to kill an organism; the more toxic the substance the smaller the lethal dose.

If a sample of aquatic animals is exposed to a toxin, not all die at the same time. Instead, mortality shows a sigmoid relationship to the period of exposure (Fig. 2.1). The **median lethal time** is the time for the death of 50 per cent of a sample and is written **LT$_{50}$** or **LT$_m$**.

The lethal time depends on the concentration of toxin to which the organism is exposed—the higher the concentration, the shorter the time—and there may be a lower threshold concentration below which the material is not toxic. The LT$_{50}$ is therefore not a very useful statistic and it is more usual to determine the concentration of toxin at which 50 per cent of the test organisms are killed within a specified time. The time is commonly 48 or 96 hours and the toxicity is then recorded as the **median lethal concentration** and written **96 h LC$_{50}$**.

The median lethal concentration is measured by determining the median lethal time

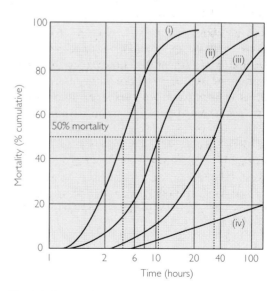

Fig. 2.1 Cumulative percentage mortality of mussels, *Mytilus* in (i) 10^{-2} M, (ii) 10^{-3} M, (iii) 10^{-5} M, (iv) 10^{-3} M solution of zinc sulphate.

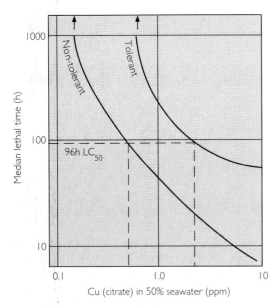

Fig. 2.2 The 96 h LC_{50} of copper determined from the relationship between copper concentration (as citrate) in 50 per cent seawater and the median lethal time for copper-tolerant (from Rostronguet Creek) and copper-intolerant (from the Avon estuary) *Nereis diversicolor*.

at different concentrations, the latter generally being at logarithmic intervals, and reading the LC_{50} from a plot of the results (Fig. 2.2).

Antagonism and synergy

In the natural environment, toxins are rarely present in isolation and they may interact with other substances. The combined effect of several toxins may be:

- the addition of one mortality to another;

- the mortality caused by one includes that caused by the other;

- one increases the mortality caused by the other (**synergy**);

- one reduces the mortality caused by the other (**antagonism**).

Table 2.1 shows the results of an experiment in which the ciliate protozoan *Cristigera* was exposed to zinc, mercury, and lead salts in various combinations and concentrations. The average percentage reduction in the growth of the culture was recorded in each case.

If the toxic effects are additive, the sum of the growth reductions for each of the metals acting alone should be the same as the reduction caused by all three acting together. In this example, 0.005 ppm mercuric chloride reduces growth by 12.1 per cent, 0.3 ppm lead nitrate causes 11.8 per cent reduction, and 0.25 ppm zinc sulphate causes 14.2 per cent reduction. The sum, 12.1 + 11.8 + 14.2 = 38.1 per cent, should be the same as the effect of these concentrations of salts in a mixture. In fact, as shown by the bottom right-hand figure in the table, it is 67.8 per cent; the combined effect is synergistic.

A synergistic effect is often noted in organisms exposed to natural environmental stress as well as to a toxin. Figure 2.3 shows how the resistance of the fiddler crab *Uca* to cadmium is affected by increased temperature and reduced salinity. The 240 h LC_{50} is about 48 ppm when the crabs are under normal conditions of temperature (10 °C) and salinity (30 per mille), but falls rapidly if the salinity is below 20 per mille. A combination of low

Table 2.1 Percentage reduction of growth of *Cristigera* caused by different concentrations and combinations of metal salts

$ZnSO_4$ (ppm)	0			0.125			0.25		
$Pb(NO_3)_2$ (ppm)	0	0.15	0.3	0	0.15	0.3	0	0.15	0.3
$HgCl_2$ (ppm)									
0	0	8.5	11.8	8.3	14.4	18.8	14.2	18.3	25.9
0.0025	9.5	10.7	14.5	13.9	16.2	22.6	18.8	29.0	51.3
0.005	12.1	18.7	21.8	18.9	21.3	23.2	35.5	36.5	67.8

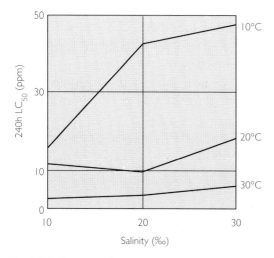

Fig. 2.3 Influence of temperature and salinity on the 240 h LC_{50} for cadmium to the fiddler crab *Uca pugilator*.

salinity (10 per mille) and high temperature (30 °C) reduces the LC_{50} to only 2–3 ppm.

In contrast to this synergistic effect, a mixture of 0.0025 ppm mercuric chloride (causing 9.5 per cent reduction in growth of the *Cristigera* colony if used alone) and 0.15 ppm lead nitrate (causing 8.5 per cent reduction) produces only 10.7 per cent reduction in the growth of the population, instead of 9.5 + 8.5

= 18.0 per cent reduction (Table 2.1) expected if the results were additive; they are antagonistic.

A variety of antagonistic effects of multiple contaminants has come to light. Well-known examples are selenium and mercury, and zinc and cadmium, the first contaminant reducing the effect of the second in each case.

Some technical problems

Designing toxicity tests is not always straightforward. The test organisms have to be kept long enough to demonstrate effects and during this time test conditions have to be maintained at a constant level. The form in which the toxin is presented to the organism may also have a profound effect on the results of a test.

This is particularly true of metals, the toxicity of which may be radically different depending on the form of the metal to which the organism is exposed. Valency may change in seawater and is influenced by salinity changes (Table 2.2), or complexes may be formed between the metal and organic substances. Figure 2.4 shows the uptake of lead by the mussel *Mytilus edulis* when the metal is added as the nitrate, citrate, or complexed with organic molecules. Metal citrates are commonly used in toxicity tests because of the

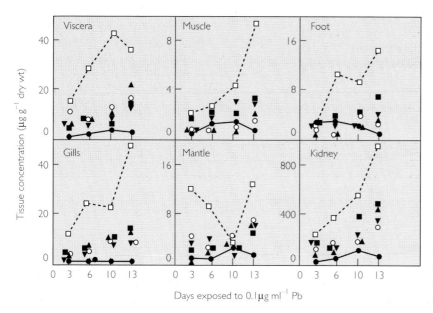

Fig. 2.4 Uptake of lead by various tissues of the mussel *Mytilus edulis*, with lead added as nitrate (○), citrate (□), humate (▲), alginate (▼), and pectinate (■), with seawater control (●). (*Elsevier Science*)

Table 2.2 Concentration and speciation of trace metals in seawater

Metal		Area	Concentration ($\mu g\,l^{-1}$)	Main species in aerated water 35% salinity	10% salinity
Silver	Ag	N.E. Pacific	0.00004–0.0025	$AgCl_2^-$	
Aluminium	Al	N.E. Atlantic	0.162–0.864	$Al(OH)_4^-$, $Al(OH)_3$	
		N. Atlantic	0.218–0.674		
Arsenic	As	Atlantic	1.27–2.10	$HAsO_4^{2-}$	
Cadmium	Cd	N. Pacific	0.015–0.118	$CdCl_2$, $CdCl_3$, $CdCl^+$	Cd^{2+}, $CdCl^+$
		Sargasso Sea	0.0002–0.033		
		Arctic	0.015–0.025		
Cobalt	Co	N.E. Pacific	0.0014–0.007	$CoCO_3$, Co^{2+}	Co^{2+},$CoCO_3$
Chromium	Cr	E. Pacific	0.057–0.234	CrO_4^{2+}, $NaCrO_4^-$	CrO_4^{2+}
Copper	Cu	Arctic	0.121–0.146	$CuCO_3$, Cu-organic	Cu-humic, $Cu(OH)_2$
		Sargasso Sea	0.076–0.108		
Iron	Fe	Arctic	0.067–0.553	$Fe(OH)_3$, $Fe(OH)_2^+$	
Mercury	Hg	N. Atlantic	0.001–0.004	$HgCl_4^{2-}$, $HgCl_3^-$	Hg-humic, $HgCl_2$
Manganese	Mn	Atlantic	0.027–0.165	Mn^{2+}, $MnCl^+$	Mn^{2+}
		Sargasso Sea	0.033–0.126		
Nickel	Ni	Arctic	0.205–0.241	$NiCO_3$, Ni^{2+}	Ni^{2+}, $NiCO_3$
		Sargasso Sea	0.135–0.334		
Lead	Pb	Central Pacific	0.001–0.014	$PbCO_3$, $PbOH^+$	
		Sargasso Sea	0.005–0.035		
Antimony	Sb	N. Pacific	0.092–0.141	$Sb(OH)_6^-$	$Sb(OH)_6^-$
Selenium	Se	Pacific and Indian	0.044–0.170	SeO_4^{2-}, SeO_3^{2-}	
Tin	Sn	N.E. Pacific	0.0003–0.0008	$SnO(OH)_3^-$	
Vanadium	V	N.E. Atlantic	0.83–1.57	HVO_4^{2-}, $H_2VO_4^-$	
Zinc	Zn	N. Pacific	0.007–0.64	Zn^{2+}, $ZnCl^+$	Zn^{2+}
		Sargasso Sea	0.004–0.098		
		Arctic	0.056–0.225		

ready solubility of this salt, but lead citrate has three or four times the final tissue concentration and rate of uptake as the nitrate. Complexes with organic molecules have double the rate of uptake. Organic complexes of heavy metals also tend to be more toxic than inorganic compounds: organomercurials (Table 2.3) and organotins are 10–100 times more toxic. Organic arsenic, on the other hand, is less toxic than inorganic arsenical compounds.

Other toxins present different problems for designing realistic experiments. Oil, for example, is immiscible with water and toxicity is usually measured for the water-soluble components; although the concentration and identity of these are not known in detail, and, in any case, in the natural environment the

Table 2.3 Toxicity of various mercury compounds to the red alga, *Plumaria elegans*

Compounds	18 h LC_{50} (ppb Hg)
Methyl mercuric chloride	44
Ethyl mercuric chloride	26
n-Propyl mercuric chloride	13
n-Butyl mercuric chloride	13
n-Amyl mercuric chloride	13
Isopropyl mercuric chloride	28
Isoamyl mercuric chloride	19
Phenyl mercuric chloride	54
Phenyl mercuric iodide	104
Mercuric iodide	156
Mercuric chloride	3120

organism is likely to ingest emulsified droplets of whole oil. Furthermore, volatile constituents of the oil evaporate so that the nature of the material to which the test organism is exposed changes throughout the course of an experiment.

Chlorinated hydrocarbon pesticides have very low solubility in water and, in nature, animals are usually exposed to these compounds when they are adsorbed on to particles that the animal ingests. This is impossible to simulate accurately in the laboratory because of sedimentation of the particles to the bottom of the container, and adsorption of the chlorinated hydrocarbons on to the walls of the test chamber, which progressively reduces the availability of the material to the test organisms.

Organisms vary widely in their sensitivity to a toxin. Different species, of course, differ in their sensitivity, but even within a single species, sensitivity to a toxin depends on age, sex, reproductive condition, exposure to other stresses, nutritional state, and previous history, as well as to the genetic constitution of the test organisms. The wide range of variation that may be caused by these factors is illustrated in Table 2.4: young forms, for example, may be 100 times less sensitive, or 1000 times more sensitive than adults of the same species.

Toxicity tests have an important but limited role in pollution studies. While they may give some guidance as to the most susceptible organisms in a community exposed to an effluent, or indicate which constituent in a mixture of wastes is probably responsible for observed damage, they do not give reliable predictions about the impact of an effluent in the natural environment.

Sublethal effects

It is often possible to detect responses in organisms to toxins at far lower concentrations than those that kill them. Sublethal responses vary widely but may include major physiological stress, tumours, or developmental abnormalities that would be likely to result in early death.

- Erosion of fins and precancerous growths (papillomas) on the ventral surface are commonly observed on flatfish living on very contaminated sediments. The response is not specific, and has been observed whether the contaminant is sewage sludge, oil, or titanium dioxide waste.

- The ingestion of crude oil by herring gulls (*Larus argentatus*) and some other sea-birds causes damage to the intestine and liver, and impairs the functioning of the nasal salt glands.

- Skeletal deformities are relatively common in fish, particularly in those from polluted waters (Fig. 2.5): as much as 8.4 per cent of herring (*Clupea harengus*) caught in the southern North Sea are afflicted in this way. Natural causes, such as hereditary factors, developmental errors, and parasitic infection are partly responsible, but a variety of toxins, including chlorinated hydrocarbons, dispersed oils, and heavy metals can also cause the condition.

- Sublethal concentrations of 0.05–0.10 ppm copper sulphate and zinc sulphate cause the production of abnormal bifurcated larvae in *Capitella capitata* (Fig. 2.6). A similar abnormality occurs in 10–16 per cent of larvae of this species in the second generation of

Table 2.4 Sources of variation in bioassays

Source of variation	Ratio for comparing responses	Ratio range
Age of test organism	Old:young	0.01 to 1000
Sex of test organism	Male:female	0.5 to 5 or wider range
Genotypic difference	More resistant genotype:less resistant genotype	1 to 100 or wider range
Acclimation regime	Acclimated:non-acclimated	1 to 10 or wider range
Duration of exposure	Short bioassay test:long bioassay test	1 to 100

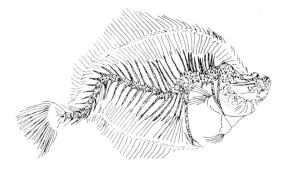

Fig. 2.5 Deformation of the skeleton of a turbot taken from polluted waters.

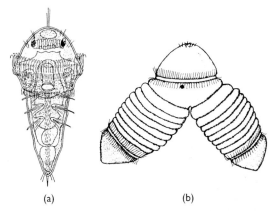

(a)   (b)

Fig. 2.6 (*a*) Normal polychaete larva, (*b*) abnormal larva of *Capitella* living in sublethal concentrations of copper or zinc sulphate. (*Elsevier Science*)

worms reared in water contaminated with detergent.

Detoxication

A number of detoxication mechanisms exist that reduce the harmful effects of substances that cannot be readily excreted.

- Exposure of animals to low concentrations of metals induces the synthesis of **metallothioneins**. These are low molecular weight proteins that form a complex with the metal so that it is then unable to become involved in chemical processes, which is the source of its toxicity. This phenomenon has been found in a variety of animals and is probably widespread. Cadmium-binding proteins have been identified in seals, rockfish (*Sebastodes*), and molluscs; copper- and cadmium-binding

thioneins occur in crabs; metallothioneins incorporating zinc, copper, and cadmium are found in the sea urchin *Strongylocentrotus purpureus*.

- A second method of detoxicating metals has been found in nearly all animal groups. The metal is stored in granular form; the **granules** are bound by a membrane and are therefore isolated from the chemical activity in the cell. The membrane may be derived from the endoplasmic reticulum, the golgi apparatus, or the cell membrane of the storage cell. So far three types of granule have been found, containing copper, iron, or calcium; the calcium granules may be pure or may contain other metals including manganese, magnesium, phosphorus, zinc, cadmium, lead, and iron.

- Lipophilic organic compounds, such as polyaromatic hydrocarbons and PCBs induce the synthesis of **mixed function oxygenases** (MFO) which are widespread in animal tissues, although most studies have related to those in the liver of fish. The majority of the oxidative reactions mediated by MFOs are attributed to cytochrome P450 enzymes. These render the lipophilic hydrocarbon more water soluble and therefore more easily excreted in the urine or bile. Unfortunately, the resulting metabolites may be carcinogenic, and liver tumours are common in fish exposed to large concentrations of organic pollutants.

Scope for growth

Cellular responses to pollutants are specific to the toxin; the '**scope for growth**' test developed for use with bivalves, especially mussels (*Mytilus*), gives a more general measure of the overall physiological health of the animal based on its energy budget. The scope for growth is the difference between the energy assimilated from food, and the energy used in respiration, excretion, and other maintenance activities. Any surplus energy is available for growth and reproduction. A reduced, or even negative, scope for growth results when the energy intake from food is reduced, as it may be in winter, or if energy expenditure on maintenance activities is increased by environmental stress. Metabolic reserves may then have to be used to support maintenance activities and, in those circumstances, growth and reproduction is impossible. Figure 2.7 shows how the scope for growth of *Mytilus edulis* is reduced in a pollution gradient in Narragansett Bay, Rhode Island.

Physiological response and survival

Molecular biomarkers, such as detoxication agents (metallothioneins, mixed function oxy-

genases, etc.) in an animal, or a low or negative scope for growth in a mussel indicate that the animals are, or have been, under stress. Furthermore, the concentrations of biomarkers can be measured to give an indication of the degree of exposure of toxins.

These physiological responses to exposure to stress, including exposure to toxins, are often harmful, but not inevitably so.

• The polychaete *Nereis diversicolor* living in an estuary heavily contaminated with copper

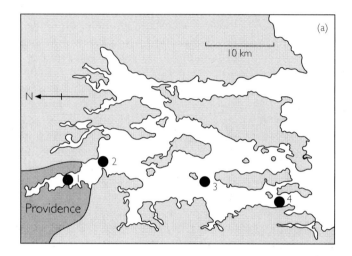

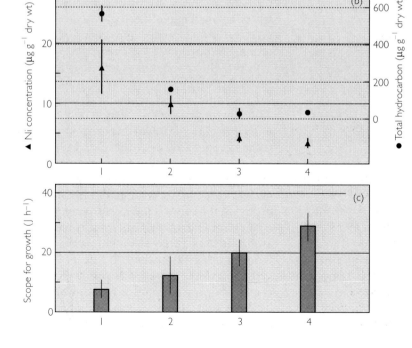

Fig. 2.7 Scope for growth of *Mytilus* (c) at stations in a pollution gradient (b) in Narragansett Bay (a). (*Elsevier Science*)

has evolved a copper-tolerant genotype (Fig. 2.2, p. 12).

• Commercial fish hatcheries often subject fry to as many as possible of the stresses they are likely to encounter in nature to enhance survival.

In theory, such adaptive physiological responses are distinguished from harmful ones if they contribute to the survival, growth, and reproduction of the species, but the distinction is not always easy to make.

POPULATION CHANGE

From a biological point of view, when pollution causes the death of organisms, what matters is not the initial mortality, but the numbers and fate of the survivors. Mortality that results in a prolonged reduction in the population of a species is obviously regarded more seriously than a loss that is rapidly made good. Population changes in particular species, of course, have some impact on the community of which they are a part, and the pollution impact may be measured at the level of populations or communities.

Key species

Although many species may be affected, attention is often focused on only a few key species:

• those of high conservation interest (some sea-birds, seals, cetaceans);

• species of commercial value and under cultivation (caged salmon, shellfish beds);

• key species whose presence or absence in certain environments has major repercussions in the community (dominant herbivores, such as limpets, on rocky substrata);

• indicator species, known to be particularly resistant (such as *Capitella*) or particularly sensitive to pollutants; the presence or absence of these indicators may provide a warning signal of the existence of pollution effects.

Measurement of population change

The abundance of a species is measured by population density or biomass. Whatever parameter is measured, it is necessary to demonstrate that a change is related to pollution by comparison with controls. This may be by comparing the population density at the same site before and after the introduction of a supposed pollutant, or by using a comparable but unaffected area as a control site. A preliminary **'power analysis'** is needed to determine how many replicate samples will be required to identify a change within acceptable statistical limits. If there is great variation between replicate samples, an inordinate number of them may be needed to establish reliably that a change has taken place. In that event, a different monitoring strategy is required. Even if changes are detected, additional investigations may be necessary to confirm that the changes are caused by the pollutant and not some other factor, such as climate change.

COMMUNITY RESPONSE

A more realistic approach than the study of the fate of selected species is to examine the response of the whole community; that is, the whole assemblage of species in an area. This is now the most popular approach in pollution impact studies.

Generally, community studies involve taking samples of organisms from polluted and control sites, identifying and enumerating the species, and then analysing the resulting data to determine if significant changes have occurred in the stressed system. A variety of components of the biota are available for community studies (fish, plankton, algae, and so on), but soft-sediment invertebrates are most commonly used and are probably the most suitable because their lack of mobility means that any observed changes are likely to be owing to pollution stress rather than migration or movement.

Identification of all the species is time-consuming, but for the purposes of detecting pollution impact, identification to a higher taxon (family, order, phylum) may be adequate.

The data may be analysed by univariate, graphical, or multivariate methods.

Univariate analysis

A single numerical index is calculated to characterize the community. Generally, indices are calculated for each replicate sample, allowing statistical comparison of means between polluted and control sites. These indices are usually an expression of **diversity**, which may be simply the number of species present per unit area (**species richness**), or a more complex quantity such as the **Shannon–Wiener** diversity index, which integrates the number of species with their abundance in the community. This index usually ranges between 0 and >3, with low values indicating stressed conditions as a result of the low number of species present and/or dominance by one particular, robust species. Other indices, measuring 'evenness' or 'dominance', are also used.

Graphical methods

There are several methods of giving a visual representation of the community.

• Species are ranked in order of percentage abundance (that is, the percentage of individuals in the sample belonging to each species) and the number of species is plotted against rank (Fig. 2.8(b)). Impacted sites dominated by one or two species have a much shallower curve than unimpacted sites where no dominance is evident.

A plot of the cumulative percentage against species rank gives a **k-dominance curve**, which, similarly, is shallower for impacted than for unimpacted sites (Fig. 2.8(c)).

• A comparable plot of cumulative biomass is often superimposed on the k-dominance curve to give the **Abundance–Biomass Comparison (ABC) Curve** (Fig. 2.9). Polluted sites are indicated by the *k*-dominance curve lying

(a)

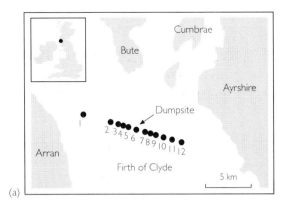

(b)

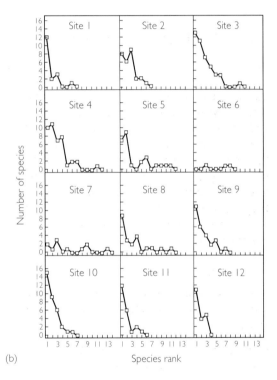

(c)

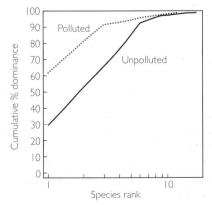

Fig. 2.8 Species abundance plots (*b*) of the benthic macrofauna at stations on a pollution gradient at the sewage sludge dump site in the Firth of Clyde (*a*), and (*c*) the k-dominance curve. (See also Fig. 2.12.)

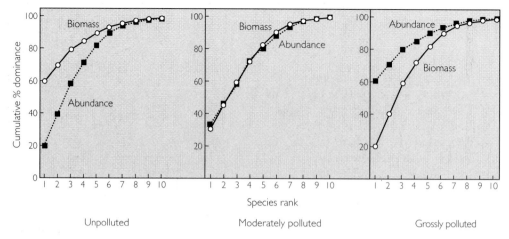

Fig. 2.9 Hypothetical ABC curves for unpolluted, moderately polluted, and grossly polluted conditions.

above the biomass curve across the entire range. Differences between the curves can be tested statistically. ABC curves are a useful tool in pollution assessment because they do not need information about the status of the site before a pollution incident.

• Other graphical techniques include **rare-faction** curves (comparing the number of species recorded from a given number of individuals encountered) and **log-normal** plots (assessing the deviation of curves from a straight line when abundances of species are plotted on a geometric scale).

Multivariate analysis

Graphical techniques keep the species separate but their identity is not important, so two sites with totally different species present could theoretically produce identical ABC plots and diversity indices, if the patterns of abundances in the two communities were similar. Important information on the actual species comprising the community is retained by multivariate statistical techniques. The aim of multivariate analysis is to determine how closely related the sites are in their species composition in order to detect any divergence from the control community structure. The analysis involves a great deal of repetitive calculation and computers are essential. Three methods are in common use.

Cluster analysis uses similarities between samples and groups of samples to build a dendrogram (Fig. 2.10). Samples with similar structures (distribution of individuals between species) are linked in pairs and plotted on a scale which reflects the similarity between the members of a pair. The pairs are then linked in second-order pairs, and so on. The subsequent levels of pairing reflect the similarity between groups less and less accurately and, in fact, different methods of analysis often give very different dendrograms for the same data. Depending on the similarity coefficient used, it may be possible to attach a level of statistical significance to the delimitation of groups of samples.

Figure 2.10 shows an index of the metal contamination at a series of sites near the Huelva estuary on the Atlantic coast of Spain. Cluster analysis of the data shows that the sites can be assigned to four groups relating to the degree of metal contamination. Group A is the least contaminated and stations 9 and 10 are the most contaminated.

Non-metric Multidimensional Scaling (MDS) starts from the calculation of similarities between samples, but, unlike a dendrogram, does not proceed beyond the first level of comparison between them. In order to represent the relationship between all the samples, it plots them so that the rank order of the similarities between pairs of samples matches

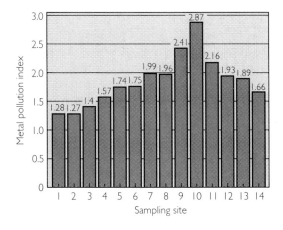

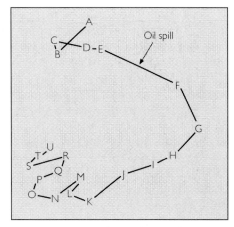

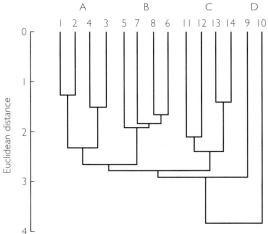

Fig. 2.10 Cluster analysis dendrogram of sites near the Huelva estuary in Spain, where there is metal contamination. (*a*) A composite index of metal contamination at the twelve sites; (*b*) dendrogram showing the similarity between sites that fall into four groups; A, least contaminated and D, most contaminated.

Fig. 2.11 MDS ordinations of macrofauna community data at various times (A–U) before and after the *Amoco Cadiz* oil spill.

also clear that the stable community following recovery (points K–U) was different from the assemblage present before the spill (points A–E).

Principal Component Analysis also produces a two- or three-dimensional plot of the samples (Fig. 2.12). The axes of the plot are defined in such a way that the Principal Component 1 represents as much as possible of the variation in the data set. It can be envisaged as some sort of 'line of best fit' through the

the rank order distance between that pair of samples in the plot. The plot may be produced in two or three dimensions, but as the rank distances are used, the axes have no units. These 'maps' provide valuable and easily interpreted information on environmental change.

Figure 2.11 shows the MDS ordination of macrobenthic community data before and after the *Amoco Cadiz* oil spill. Points A–U represent samples taken at successive times. There was clearly a dramatic change in the community structure after the oil spill, but it is

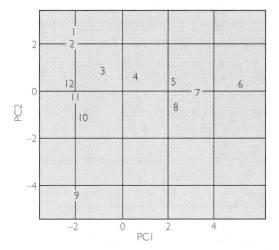

Fig. 2.12 Principal coordinate analysis of the environmental variables at twelve stations across the sewage sludge dump site shown in Fig. 2.8(*a*). (*Springer-Verlag*)

data points. Principal Component 2 is then computed to explain as much as possible of the residual variation in the data (that is, variation not explained in Principal Component 1). For a complete analysis, this process is continued through successive principal components, each explaining as much as possible of the variation remaining after the preceding operation, until all the variation has been explained. As a relationship exists between the original data and the principal components, they have units, and it is possible to calculate the correlation between them and the original variables.

There are problems in the use of PCA on data sets containing many zero values, which is typical of biological data, but it is very powerful when applied to environmental data. For example, data on the sampling depth, silt content, and levels of carbon and various metals at sites across the Garroch Head sewage sludge dumping grounds in the Firth of Clyde (see Fig. 2.8(a)) have been analysed by PCA (Fig. 2.12). In this case, PC1 explains 61 per cent of the variance in the data matrix and broadly represents increasing contaminant levels. PC2 explains a further 17 per cent; thus the two-dimensional plot represents 88 per cent of the variation in the data set. Reading along the PC1 axis, there is a clear trend from one end of the transect at stations 1 and 2, through the centre of the transect at station 6, and back to relatively clean stations 10, 11, and 12 at the other end of the transect.

It may also be possible to relate the PCA scores to other data; for example, a correlation between PC1 and the abundance of a particular species would indicate that it was metal tolerant.

OXYGEN-DEMANDING WASTES

The amount of oxygen dissolved in the water has a profound effect on the plants and animals living in it. Wastes that directly or indirectly affect the oxygen concentration are therefore of great importance.

OXYGEN DEMAND

By far the greatest volume of waste discharged to watercourses, estuaries, and the sea is sewage, which is primarily organic in nature and subject to bacterial decay. Bacterial degradation results in the oxidation of organic molecules to stable inorganic compounds.

- **Aerobic bacteria** make use of oxygen dissolved in the water to achieve this.

$$C_6H_{12}O_6 + 6O_2 \rightarrow 6H_2O + 6CO_2$$
(glucose) (oxygen) (water) (carbon dioxide)

As a result of this bacterial activity, the oxygen concentration in the water is reduced, but this is compensated for by the uptake of atmospheric oxygen. However, oxygen diffuses only slowly through water and there is a time-lag before the oxygen used by bacterial activity is replenished.

- If the oxygen concentration falls below about $1.5 \, \text{mg l}^{-1}$, the rate of aerobic oxidation is reduced. **Anaerobic bacteria** can oxidize organic molecules without the use of oxygen, but the end-products include compounds such as H_2S (hydrogen sulphide), NH_4 (ammonia), and CH_4 (methane) which are toxic to many organisms, and this process is much slower than aerobic degradation. There is therefore a

likelihood of waste accumulation.

- Some **inorganic wastes** become oxidized in water without the intervention of bacteria and these, too, deplete the water of oxygen.

Measurement of oxygen demand

When planning the discharge of any waste to water, it is important to know the amount of oxygen required to degrade it (the **oxygen demand**) in order to avoid undesirable environmental consequences. The chemical composition of nearly all organic wastes is extremely complicated and different constituents require different amounts of oxygen to achieve complete oxidation. It is impracticable to analyse a type of waste to discover its exact content, so, in order to measure the effluent load it represents, the overall oxygen demand for complete oxidation is measured directly.

Chemical oxygen demand (COD) is measured by adding an oxidant such as potassium permanganate ($KMnO_4$) or potassium dichromate ($K_2Cr_2O_7$), with sulphuric acid (H_2SO_4), to a sample of effluent. The sample is titrated after a standard interval to determine the amount of oxidant remaining. From this, the total amount of oxidizable material can be calculated.

Biochemical oxygen demand (BOD) is the usual method of measuring the oxygen demand of organic wastes. The oxygen concentration in a sample is measured before and after bacterial digestion for a standard time (for example three or five days, recorded as BOD_3 or BOD_5, respectively). It may be necessary to add bacteria and nitrate if these are deficient in the initial sample. This gives a direct measure of the amount of oxygen used in the bacterial degradation of the sample.

THE DILUTION FACTOR

If water is saturated with oxygen, there is sufficient oxygen to oxidize a BOD_5 of about 8.0–8.5 mg l^{-1}, but the BOD_5 of organic effluents is usually much greater than this. Urban sewage commonly has a BOD_5 of 500 mg l^{-1}; spilled beer has a BOD_5 of 70000 mg l^{-1}. It is therefore necessary to find some way of diluting the effluent to achieve a BOD_5 of about 8 mg l^{-1}, hence the old adage of sanitary engineers: 'the solution to pollution is dilution'.

The common practice is to discharge the effluent into a large volume of water, preferably a river or the sea, where water movement produces mixing and achieves the necessary dilution. If the effluent is discharged into a river (Fig. 3.1), it is swept downstream and mixes with the river water in a **mixing zone**.

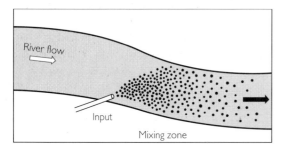

Fig. 3.1 Discharge of an organic effluent into a river, and its mixing zone.

The concentration of effluent is high near the outfall but progressively reduced downstream. The dilution achieved depends upon the rate of flow of the river, its organic load, and the rate of input of the effluent and its oxygen demand.

For example,

River flow   8 m³s^{-1} with BOD 2 mg l^{-1}
Effluent input   1 m³s^{-1} with BOD 20 mg l^{-1}

$$\text{BOD after mixing} = \frac{\text{total BOD}}{\text{total volume}}$$

$$= \frac{(8 \times 2) + (1 \times 20)}{8 + 1} = 4$$

This achieves the necessary dilution.

OXYGEN BUDGET

Bacterial activity consumes oxygen. Figure 3.2 shows the effect of an organic input on the dissolved oxygen in the water, either with time, following a single input to a static body of water, or with distance downstream from a continuous input to flowing water.

As the bacteria multiply, oxygen consumption increases to a peak and then declines as the organic matter is oxidized. This extracts oxygen from the water column and, when the dissolved oxygen falls below saturation, oxygen is taken up from the atmosphere; the greater the oxygen depletion of the water, the greater the rate of oxygen uptake. Uptake lags behind utilization because of the time taken for oxygen to diffuse through the water.

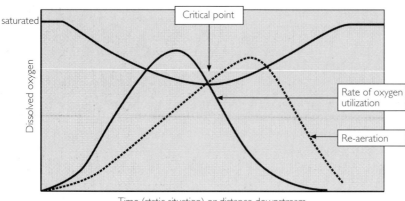

Fig. 3.2 Oxygen budget for a river or a body of water receiving organic wastes. The horizontal axis represents time for a single input of waste to a static body of water, or distance downstream from a continuous input to a river.

At a **critical point**, the rate of reaeration overtakes the rate of oxygen utilization and this corresponds to the minimum concentration of dissolved oxygen in the water.

Thus, the effect of an organic input is to produce an **oxygen sag**, corresponding to the dissolved oxygen in the receiving water, less bacterial respiration, plus reaeration. The oxygen sag may be trivial, but if the receiving waters are seriously overloaded with organic material it may be sufficient to produce anoxic conditions.

ESTUARIES

The situation in tidal waters is much more complicated than that in rivers.

Settlement

The previous calculations (p. 24) assumed that all the discharge is in solution or suspension, but in practice the effluent often contains solids that settle to the bottom, and it is important to know the settlement characteristics of the discharge if the dilution factor is to be calculated properly. The rate of settlement of particles in water depends on their size and density, and the viscosity and speed of the water; it is governed by Stokes' law and Newton's law. However, particles tend to flocculate into larger aggregations and Stokes' and Newton's laws do not apply in these circumstances.

In estuaries, the change of pH and redox potential as freshwater encounters the sea causes intense flocculation of clay and other particles, with increased adsorption of metals and other materials on the flocculates. While sedimentation takes place in all but the swiftest rivers, estuaries are subject to particularly heavy sedimentation, leading to the development of extensive mud-flats containing much of the organic material, metals, pesticides, and so on, from the water column.

No two situations are the same and a detailed study of the settlement characteristics of an effluent in the receiving waters is needed before it can be calculated how much,

where, and at what rate waste can be discharged, to reduce the BOD to within the capacity of the receiving waters.

Residence time

Whereas rivers flow in one direction, there is a tidal ebb and flow in estuaries and the net seaward flow over the complete tidal cycle may be small. Mean figures for the Thames at London are:

high water to low water:
downstream flow 15 km

low water to high water:
upstream flow 13 km

giving a net seaward flow of only 2 km per tidal cycle. Towards the seaward end of the estuary, the seaward flow becomes very slow indeed (Fig. 3.3).

Thus, a body of water (or effluent) entering the estuary moves through it only slowly and has a long **residence time**. For example, it takes about 30 days for water at the head of the Thames estuary to reach the sea. It follows that the dilution capacity of the estuary is correspondingly reduced. The oxygen sag is more pronounced and may create an **anoxic zone** from which most life disappears, except for anaerobic bacteria, fungi and yeasts, and some Protozoa. The estuary becomes foul smelling from the products of anaerobic oxidation, and migratory fish such as salmon—or more

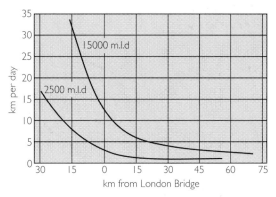

Fig. 3.3 Net seaward flow of water in the Thames estuary at two rates of river flow in million litres per day. (*Published with the permission of the Controller of Her Majesty's Stationery Office.*)

particularly their young, which require a minimum oxygen concentration of 4 mg l^{-1}—cannot pass through this zone.

Stratification

Depending on their shape, some estuaries, such as the Tees, show **stratification**, with the denser seawater flowing upstream in a wedge, and the river water flowing out over it (Fig. 3.4). Other estuaries, such as the Thames, are not stratified. If an effluent is discharged into the bottom water layer of a stratified estuary and has neutral buoyancy, it may actually move upstream until it reaches the end of the wedge and comes under the influence of the seaward-moving upper layers of water. If it has negative buoyancy, it may never travel seaward at all.

Mixing zone

In rivers, there is a defined mixing zone downstream of the point of effluent discharge (Fig. 3.1). In estuaries, the mixing zone is downstream of the point of discharge during a falling tide; but on a rising tide the mixing zone may swing upstream from the point of discharge and, as a result a considerable stretch of the estuary comes under the influence of the effluent. Particularly in stratified estuaries, where different effluents behave differently, the whole estuary constitutes the mixing zone.

SEWAGE TREATMENT

River water with a BOD_5 of less than 2 mg l^{-1} can be regarded as unpolluted; with a BOD_5 greater than 10 mg l^{-1} it is grossly polluted. Between these extremes, various standards may be set, depending on the purpose for which the river water is required. Water for salmon or trout should have a BOD_5 less than 3 mg l^{-1}, for coarse fish (that is, other than salmonids) less than 6 mg l^{-1}. Drinking water may have a BOD_5 up to 7 mg l^{-1}, but drinking water quality is determined by many other factors including nitrate, heavy metal, pesticide, and bacterial content. Irrigation water may have a very high BOD_5.

A common dilution factor in rivers is 8:1. To achieve a final BOD_5 of 4 mg l^{-1}, which might be acceptable for many purposes, the BOD_5 of the waste should not exceed about 20 mg l^{-1} (see example on p. 24). If more waste has to be disposed of than the receiving waters can accept, and it is impossible to get the required dilution, it is necessary to introduce sewage treatment in which favourable conditions are provided for accelerated bacterial degradation of part of the organic

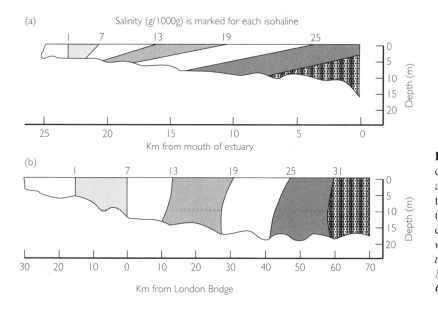

Fig. 3.4 Vertical distribution of salinity at high water of spring tides in (*a*) the Tees and (*b*) the Thames estuaries. (*Published with the permission of the Controller of Her Majesty's Stationery Office.*)

matter. This reduces the BOD of the final product before it is discharged to the receiving waters. Various stages of treatment may be used, depending on the quality of effluent that is required (Fig. 3.5).

Primary treatment

Urban sewerage contains more than 95 per cent water, and, in addition to wastes of human origin, it contains domestic and industrial waste water, solids, litter, and runoff from the land. As a **preliminary treatment**, it is screened to remove large solids, comminuted or macerated to produce a slurry, and grit is extracted.

For **primary treatment** the remainder is placed in settlement tanks. The supernatant liquid, which still has a high BOD and has a certain proportion of suspended solids, is discharged to the receiving waters. The sludge that has settled out is disposed of elsewhere.

Secondary treatment

If a greater reduction of BOD is required, the liquid is subjected to biological treatment. It may be filtered through beds of 5–15 cm rocks or coke, providing a large surface area for bacteria that degrade organic matter in the water as it percolates through the bed. After settlement, the supernatant liquid is discharged to the receiving waters and the settled material disposed of.

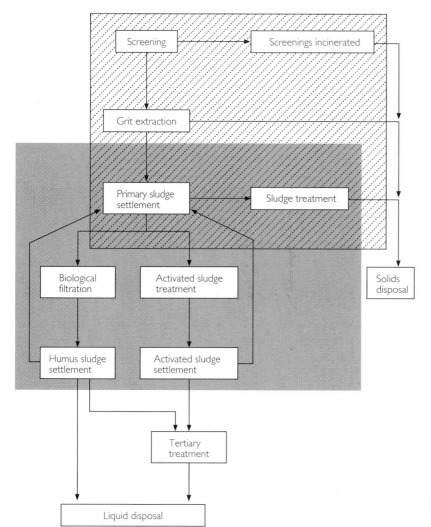

Fig. 3.5 Scheme of primary (cross-hatched) and secondary (shaded) treatment of sewage.

The sludge from the first stages of digestion gains a very high bacterial count (**activated sludge**). This may be retained, and air blown through a suspension of the effluent and activated sludge; very high rates of bacterial activity can be achieved by this method.

Tertiary treatment

If a high quality of effluent is required, it may be necessary to give further treatment. The liquid derived from secondary treatment may be retained in sedimentation ponds or passed through sand or earth filters to remove suspended solids. To reduce plant nutrients, nitrates may be removed by algal growth in retaining ponds, and phosphates can be removed by electrolytic methods.

Disinfection treatment

Some form of disinfection is increasingly used to remove pathogens such as bacteria and viruses if the human population is likely to be exposed to the effluent that is discharged, as in bathing water. This can be achieved by the addition of strongly reactive chemicals such as chlorine, ozone, or perchloric acid, but these may react with constituents of the effluent to yield potentially carcinogenic organochlorines. A safer alternative is to subject the waste to ultraviolet light or continuous microfiltration treatment, both of which remove viruses as well as bacteria. These procedures require an effluent that has received at least full secondary treatment beforehand.

Costs

The provision of sewage treatment plants is costly: a modern treatment plant giving full secondary treatment and ultraviolet disinfection to sewage for a summer population of 8000 at a seaside resort in south-west England cost £10 million in 1995. Providing the sewers and so on, costs two or three time as much again, and the annual running costs are about 5 per cent of the capital costs. In view of this, no more sewage is treated than is necessary, and advantage is taken of the natural capacity of receiving waters to degrade wastes, so far as that can be done safely.

CONSEQUENCES OF ORGANIC DISCHARGES TO ESTUARIES

The most serious problems of waste disposal are encountered in industrialized and urbanized estuaries. The long history of pollution and recovery of the Thames estuary provides an ideal case study.

The Thames was an important salmon river until about 1820. Sewage from London was either carried to fields in the surrounding countryside and used as fertilizer, or simply left to rot in cesspits or the streets. The population was largely dependent on wells and the river for drinking water; both were frequently contaminated and typhoid was common and cholera epidemics occurred at intervals. Then came an improved piped water supply, the introduction of the water closet, and the construction of a system of sewers began, discharging the wastes into the Thames. As a result of these public health measures, disease was reduced but the river was progressively overloaded and by the 1850s had become foul smelling and devoid of fish.

In the later years of the nineteenth century, the number of separate outfalls into the river was reduced and the sewage piped to settlement plants; the resulting sludge was dumped at sea. With this reduction in the organic load on the estuary, there was a marked improvement in water quality by the end of the nineteenth century. The rapid growth of London's population in the early twentieth century outran the capacity of the sewage treatment system and by the 1950s, 30 km of the Thames through London was anoxic in dry weather (Fig. 3.6) and banks of sewage sediment were exposed at low tide. A survey in 1957 found no fish living in the Thames for a distance of 70 km through London, from Kew to Gravesend.

Improvements and expansion of the sewage treatment system between 1964 and 1976 caused a dramatic rise in dissolved oxygen levels (Fig. 3.6) and the return of a diverse benthic invertebrate fauna and of estuarine fish, and shore birds and waders feeding on the mud flats in winter. A large restocking

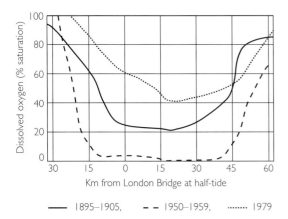

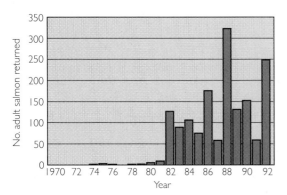

Fig. 3.6 Oxygen sag in the tidal River Thames for periods since 1895.

Fig. 3.7 Number of salmon returning to the River Thames following the restocking programme.

programme in the 1980s resulted in salmon returning in increasing numbers (Fig. 3.7).

CONSEQUENCES OF SLUDGE DUMPING AT SEA

Although in some cases the practice has now stopped, it has been common to discharge sewage sludge to sea where it was practicable to do so, either by pipeline or by dumping from barges. The effect of this input depends on the hydrography at the disposal site.

Dispersive and cumulative sites

Two of the largest sludge disposal sites in Europe were in the Barrow Deep in the outer Thames estuary (Fig. 3.8) and off Garroch

Head in the Firth of Clyde (Fig. 3.9). Neither site is now in use. The Barrow Deep received about 5 million t year^{-1} since the end of the nineteenth century, but the dump site is an area of strong bottom currents and the wastes

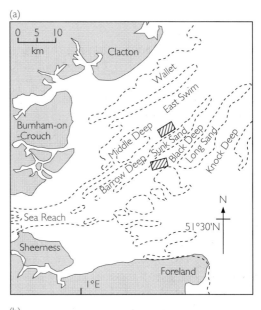

Fig. 3.8 Sewage sludge dumping in the outer Thames estuary: (*a*) location of the dumping ground, (b) organic content of sediments. (*Elsevier Science*)

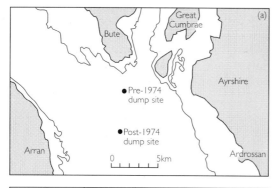

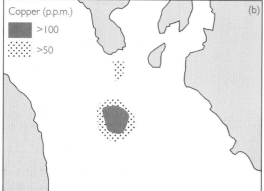

Fig. 3.9 Sewage sludge dumping in the Firth of Clyde: (*a*) location of the dumping grounds before and after 1974, (*b*) copper content of sediments in 1980.

are dispersed, having virtually no local effect, although they add to the total organic load in the southern North Sea. Garroch Head is an area of weak currents and received 1.5 million t year^{-1} of sludge; the wastes accumulated there, with a severe but localized impact on the seabed. It was a **'sacrificial' site** exactly comparable to a waste tip on land. The two sites, which represent opposite extremes, are compared in Table 3.1.

New York Bight

From 1924 until 1986, sewage sludge from New York's twenty sewage treatment plants was dumped at the '12-mile site' in New York Bight (Fig. 3.10(a)) where the water is about 50 m deep; 10 million t year^{-1} of sludge was disposed of there. The impact was intermediate between that in the Firth of Clyde and that in the dispersive grounds of the Thames estuary. There was organic enrichment and accumulation of metals with faunistic impov-

erishment in the centre of the dumping area, but there is a degree of dispersion and evidence of the spread of contaminated sediments into the Hudson Shelf Valley (Fig. 3.10(b)), and, despite the very much greater volume of wastes dumped, the concentrations of metals are less than two-thirds of those found in the Clyde. (In 1971, the maximum concentration of copper found in the Clyde dump site was 208 ppm, at the 12-mile site in New York Bight it was 132 ppm.)

From 1986 to 1991, the waste was taken to the '106-mile site' instead, 185 km offshore and with a water depth of 2200–2700 m. This is a dynamic oceanic environment and a 1991 survey of the distribution of *Clostridium* spores, which are a good indicator of sewage sludge, showed that the waste was dispersed over an area of 10 000 km^2 (Fig. 3.10(c)). The highest concentration of spores, close to the dumping zone, was 8000 g^{-1} dry sediment, compared with ten times that at the old 12-mile dump site. There were elevated concentrations in the Hudson Shelf Valley, presumably derived from slumping of sediments from the 12-mile site. Otherwise, there was no evidence that the sludge collected on the continental slope and certainly not on the shelf where most other commercial activities take place. Despite this, sea disposal of waste at the site was prohibited after 1991.

Effect on benthic organisms

Sewage and sewage sludge dumped in the sea has a direct effect on the benthic fauna, which is exposed to the sedimentation of particulate matter rich in nutrients and bacteria. Smothering by the particulate matter and reduction of the oxygen concentration because of enhanced bacterial activity excludes the more sensitive species, but the more tolerant species flourish because of the input of extra nutrients to the system. The result is a reduction in diversity but an increase in the abundance of organisms (Fig. 3.11).

The polychaete *Capitella capitata* is such an **opportunist species**, which benefits from organic enrichment. It is tolerant of chemical stresses and has a short generation time so

Table 3.1 Comparison of sewage sludge dumping grounds in the Thames estuary and Firth of Clyde

	Barrow Deep (Fig. 3.8)	Garroch Head (Fig. 3.9)
City served	London	Glasgow
Amount of sludge dumped	5 million t yr^{-1}	1.5 million t yr^{-1}
Period of use	Since end of 19th century	Since 1974 (pre 1974 dump site 6 km to north)
Bottom currents at site	Strong	Weak
Contamination of sea bed by organic waste	Wastes dispersed over a wide area, no localized enrichment evident (Fig. 3.8b)	Wastes accumulate, high levels of organic enrichment in sediments at dump site. Bottom water does not become anoxic
Contamination of sea bed by heavy metals	No elevated metal levels in sediments at dump site. Elevated Hg levels in biota in 1970s now fallen to background levels	High levels of metals at dumpsite (Fig. 3.9b), elevated levels still apparent at pre-1974 site. Fish and shellfish do not show elevated concentrations
Condition of bottom fauna	Benthic community similar to surrounding areas	Benthic community dominated by species indicative of organic pollution (such as *Capitella capitata*)

that under suitable circumstances, and in the absence of competition, its population increases rapidly. It, therefore, becomes dominant at the discharge site, but farther away, where the organic load is less and conditions are more favourable, other species become established. These out-compete *Capitella* and there is a succession of species away from the source of the waste. Figure 4.12 (see p. 56) shows a similar succession of benthic fauna in relation to oil pollution.

Species such as *Capitella* have become **indicator species** for organic pollution.

ENRICHMENT AND EUTROPHICATION

Enrichment

Many of the wastes entering the sea are plant nutrients. Decaying organic matter and nitrates and phosphates in sewage enhance plant growth. Another important source of plant nutrients is agricultural fertilizers in runoff from areas of intensive farming. Inputs of plant nutrients enhance the growth of phy-toplankton and fixed plants, and this **enrichment** benefits a number of food chains, as do the increased bacterial populations stimulated by an organic input.

A moderate input of nutrients has the same effect in the sea as adding fertilizer or manure to a garden or farm land, as the following example shows. The Seto Sea is the largest enclosed body of coastal water in Japan and provides 5 per cent of the country's commercial catch of fish. As the Japanese economy grew after the Second World War, increasing development resulted in elevated inputs of plant nutrients, boosting primary production in the Seto Sea. As a consequence, fishery catches in the sea have increased in parallel with the rise in primary production (Fig. 3.12).

Eutrophication

Although a moderate input of organic material may be beneficial, over-fertilization results in extravagant growth of plants, and the bacterial decay of dead plant material may result in oxygen depletion. Both excessive plant growth and oxygen depletion lead to alteration of the community structure, some-

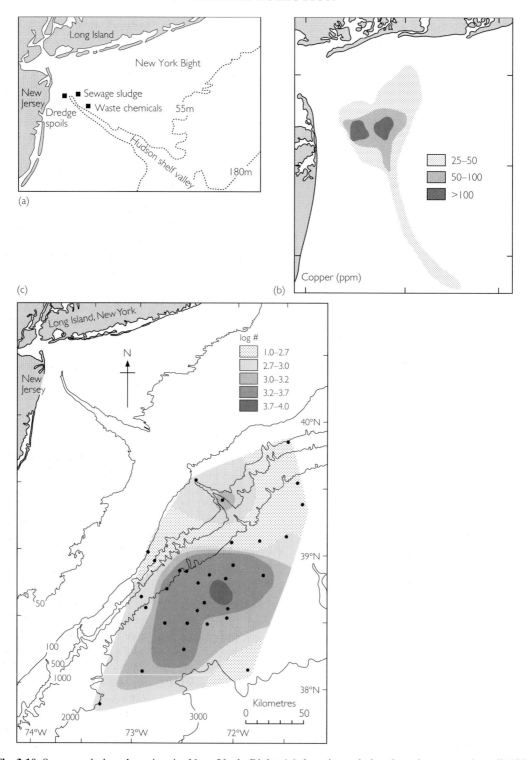

Fig. 3.10 Sewage sludge dumping in New York Bight: (*a*) location of the dumping ground until 1986, (*b*) copper content of sediments in 1972, (*c*) distribution of *Clostridium* spores at the 106-mile site in 1991. (*Elsevier Science*)

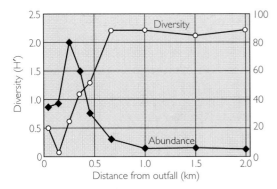

Fig. 3.11 Effect of sewage on abundance and diversity of the benthic fauna in Kiel Bay. (*Elsevier Science*)

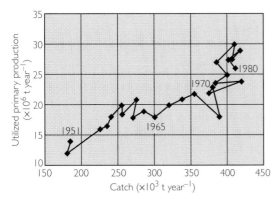

Fig. 3.12 Relationship between fish catch and primary production between 1951 and 1980 in the Seto Sea, Japan. (*Elsevier Science*)

times disastrously. These phenomena are features of **eutrophication** and are a familiar problem in freshwaters; it is only relatively recently that this has been recognized as a problem in the sea. The problems are very varied.

• A common sign of sewage pollution on the shore is the growth of green algae such as *Enteromorpha* and *Ulva*, and in some areas these form a dense covering on intertidal mud banks causing deoxygenation of the underlying sediment.

• Eutrophication can be a particular problem when the receiving water is naturally low in nutrients (**oligotrophic**). Fjords and sea lochs in high latitude regions such as Scandinavia, Canada, and Scotland are oligotrophic and salmon farming is increasingly popular there.

The faeces of the large numbers of fish and the uneaten food leads to deoxygenation of the water beneath the fish enclosures, at some risk to the salmon which require well oxygenated water. To avoid this problem, Norwegian salmon farms are located in at least 50 m depth of water; in Scotland, the practice is to move the location of the fish enclosures at intervals. It has even been suggested that salmon farming in Swedish waters is ecologically and economically unsustainable.

• Whilst many corals do very well in turbid waters, most thrive in clear, relatively oligotrophic water and there is increasing evidence that eutrophication is responsible for damage to coral reefs. Increased algal growth reduces light intensity, affecting photosynthesis by the coral's zooxanthellae, and it increases sedimentation, smothering the corals. In the late 1970s sewage discharges into Hawaiian coastal waters resulted in the rapid growth of the alga *Dictyosphaeria* which smothered corals. Following diversion of the effluents into deeper waters, coral cover gradually increased.

Algal blooms

A large input of plant nutrients often results in the development of **red tides**. These are phytoplankton blooms of such intensity (e.g. 50 million cells l^{-1}) that the sea is discoloured (not always red; the blooms may be white, yellow, or brown). Many animals, including commercially important fish species, are killed or excluded from the area, either because of clogging of the gills or other structures, or because of the toxic properties of the phytoplanktonic organisms. In 1981 and 1985, phytoplankton blooms resulted in mass kills of fish and invertebrates in Danish waters and such events have occurred repeatedly in many parts of the world.

Phytoplankton growth depends on the availability of essential nutrients (phosphorus, nitrogen, silicon), known as **limiting nutrients**, and unless there are sufficient levels of each of these, blooms will not develop, regardless of the concentration of other minerals. The construction of an inland dam in 1977 altered the input of nutrients to Dutch coastal waters changing them from a phosphorus-limited system to a nitrogen-limited system. The consequence was a series of blooms of *Phaeocystis* in subsequent years (Fig. 3.13).

Blooms of *Phaeocystis* form an unsightly brown foam which, when stranded on the shore, can be mistaken for sewage pollution. It regularly affects the coastlines of northern France, Belgium, the Netherlands, and Great Britain. A *Phaeocystis* bloom in the upper Adriatic during the summer of 1990 had a devastating economic effect on resorts in the area, which were deserted by the tourists because of the disgusting appearance of the beaches.

Danish waters in the Kattegat, as well as on its North Sea coasts are shallow and tend to become eutrophic in most summers. Following a succession of damaging phytoplankton blooms in the 1980s, Denmark introduced controls on the use of fertilizers in agriculture and on the development of coastal fish farms. Germany and the Netherlands have also introduced controls on fertilizer use.

Oxygen depletion

Another damaging effect of over-enrichment is that the dense blooms of phytoplankton result in a very high oxygen concentration in surface water because of increased photosynthesis. The abundance of decaying plant material falling to the seabed severely reduces the oxygen concentration in bottom waters and most benthic animals are killed or excluded from the area. This process is facilitated where a **thermocline** develops and the bottom water that receives the rain of decaying plant material is cut off from atmospheric oxygen by the layer of warm, less dense surface water floating above it.

Deoxygenation of bottom waters has been recorded for many sea areas.

• Togo Harbour in Hong Kong has experienced regular deoxygenation of bottom water since nutrient inputs to the harbour doubled during the 1980s. In 1988 the breakdown of a dense bloom of *Gonyaulax polygramma* resulted in the harbour bottom becoming anoxic and foul smelling. Fish and benthos were killed and fish farms in the vicinity lost 35 t of stock.

• In the spring and summer of 1976, a strong thermocline developed over a large part of the Middle Atlantic Bight (Fig. 3.14) accompanied by an intense phytoplankton bloom. Bottom oxygen concentrations were reduced over an area of 1200 km². It is estimated that 143 000 t of the clam *Spisula* were killed.

• A similar phenomenon was detected in a wide area of the German Bight and eastern North Sea during several years in the early 1980s (Fig. 3.15). In the worst affected areas almost no live fish were caught, most having left, but numerous dead flatfish (*Limanda* and *Pleuronectes*) were observed on the seabed by underwater television. Brittle stars (*Ophiura albida*), clams (*Chamelea gallina*), and no doubt other burrowing fauna were also killed. The sewage sludge dumping grounds of the city of Hamburg had previously been in this area, but this practice ceased in 1980 and the main source of plant nutrients in the area since then has been from the rivers Rhine and Elbe.

• River inputs and coastal discharges are probably responsible for the increasing eutrophication of the upper Adriatic (Fig. 3.16). Records of oxygen concentrations in bottom and surface waters since 1911 show that episodes of oxygen depletion in bottom waters first appeared in the mid-1950s off the

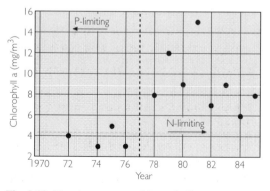

Fig. 3.13 Yearly average chlorophyll *a* concentrations off the Dutch coast. (*Springer-Verlag*)

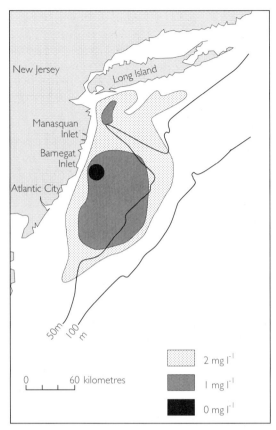

Fig. 3.14 Maximum area of reduced dissolved oxygen in bottom water of the Middle Atlantic Bight in summer 1976. The New York sludge dumping ground is in the north-west corner of the area. (*US Government Printing Office, Washington, D.C.*)

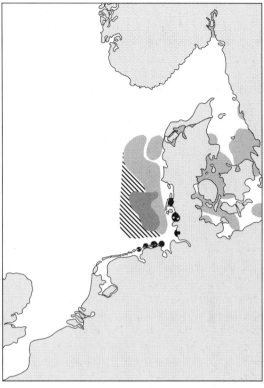

● O_2 deficiency in Wadden Sea sediments; since 1988

◼ O_2 deficiency in sediments; 1989

▨ < 2 mg O_2 l^{-1} in seawater; 1981–1984 (German Bight) and 1980–1989 (Danish marine waters)

▨ < 2 mg O_2 l^{-1} in seawater; 1981–1990

Fig. 3.15 Oxygen deficiency in seawater (<2mg l^{-1}) and sediments in the German Bight and Wadden Sea.

delta of the River Po, and that this area has extended eastwards in subsequent years. Oxygen saturation of surface waters in the same areas is associated with high rates of primary production.

● A similar seasonal deoxygenation of bottom water and supersaturation of surface water has been detected in the Gulf of Mexico, off the Mississippi Delta.

PUBLIC HEALTH RISKS

Pathogens in sewage

All human sewage contains enteric bacteria, pathogens and viruses, and the eggs of intesti-

nal parasites. Contamination of bathing water or seafood may therefore pose a health hazard to the human population:

● through contact with, or accidental ingestion of pathogenic bacteria, viruses, or yeasts;

● through infection by parasites present in the sewage or sewage sludge;

● through the consumption of contaminated seafood.

Formerly, it was believed that pathogenic bacteria and viruses did not survive in seawater and that sea bathers were at little risk from sewage contamination unless the pollution was so gross as to be visible. This is not true;

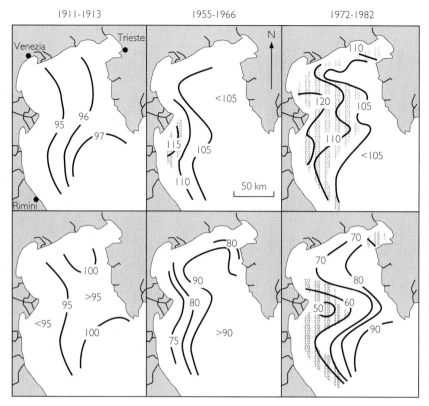

Fig. 3.16 Oxygen saturation (per cent) of surface waters (upper) and bottom waters (lower) in the upper Adriatic during the periods 1911–13, 1955–66, and 1972–82. Stippled areas experienced oxygen concentrations above 120 per cent in surface waters and below 20 per cent in bottom water. (*Elsevier Science*)

bacteria may enter a dormant phase so that they cannot be detected by normal methods (**non-platable bacteria**) and viruses can be very persistent in seawater. Thus, cuts and skin abrasions may become infected, and inadvertent swallowing of seawater may result in acute diarrhoea caused by the bacteria *Salmonella* or *Shigella*, or infection by polio or hepatitis viruses.

Parasite eggs, including those of the nematodes *Ascaris* and *Ancylostoma* (hookworm), and of tapeworms such as *Taenia*, are resistant to drying, and may persist in both crude sewage and sewage sludge. Primary sedimentation has been found to remove only 7 per cent of nematode eggs, with 66 per cent of eggs remaining viable after anaerobic digestion. The risk of transmission to humans is greatest if sewage or sludge is used on salad crops (the eggs are killed by cooking), but discharge of the waste to sea may complete the parasite's life cycle and lead to infection of humans who consume infected sea products.

The problem is greatest in tropical countries and where the incidence of intestinal parasites in the human population is high, such as parts of south-east Asia where infection rates by *Ascaris* can be 100 per cent.

The chief health risk from sewage discharges to sea is undoubtedly through the ingestion of contaminated seafood. Filter feeders, most conspicuously bivalve molluscs, flourish in waters enriched by a neighbouring input of organic wastes, but they accumulate the human pathogens on their gills and these may be transmitted to the consumer. The risk is obviously greatest if the bivalves are freshly harvested and eaten uncooked. Most bivalves in commerce that are collected from waters known to be contaminated by sewage effluents are, in fact, usually transferred to uncontaminated water for a few weeks before marketing (**depuration**), and during this time they lose their burden of pathogens and become safe to eat. Seafood organisms that are not filter feeders, such as most crustaceans and fish, do not accumulate pathogens from sewage-contaminated water and do not represent a health risk from this source.

Regulation

In the 1990 survey of the health of the world's oceans, the United Nations Group of Experts on the Scientific Aspects of Marine Pollution (GESAMP) placed sewage discharges to sea at the top of its list of concerns; heavy metal and oil pollution were given much lower priority. This was not because of the damage sewage causes in the marine environment, but because of the public health risk. In Third World countries, with poor standards of public health and low nutritional status, enteric disease transmitted through seawater is responsible for many deaths, particularly among children.

Faecal contamination is measured by a count of the bacterium *Escherichia coli* (**faecal coliform count**) or of another group of intestinal bacteria, **faecal streptococci**, which survive longer in the sea. These bacteria are always present in the human intestine and appear in the faeces. The count does not reflect the level of contamination by pathogens directly, merely the level of faecal contamination. This in itself is a good measure of the risk to which the human population is exposed, depending on the general incidence of diseases that can be communicated in this way. Bacteriological water quality standards for bathing waters and for shellfish cultivation, based on a coliform or streptococcal count, have been introduced by a number of countries. These standards are met by introducing sewage treatment and disinfection in coastal resorts, or by extending submarine outfalls so that the discharge does not return to the beaches.

Biotoxins

A considerable variety of marine organisms produce toxins that may be harmful and sometimes fatal to humans who encounter them by contact with the organisms or by ingesting them in seafood. Most such cases are unrelated to pollution, but the incidence of blooms of toxic dinoflagellates has increased in recent years, with serious public health consequences, and it is suspected that nutrient enrichment of the waters has been a contributory factor.

Ciguatera is caused by a biotoxin produced by the tropical dinoflagellate *Gambierdiscus toxicus*. The symptoms range from tingling lips and blurred vision to death from circulatory collapse, and some 50 000 cases occur annually with a fatality rate of up to 4.5 per cent. The toxins contaminate grazing fish, and human exposure is through eating predatory fish that feed on the grazers, such as snappers, groupers, and barracuda. The toxin is not destroyed by cooking. Ciguatera poisoning is predominantly found in south-east Asia and the Pacific, and has been a major constraint on the development of fisheries in some areas. Cases of ciguatera in temperate countries are a result of the consumption of imported fish and over 2000 cases per year are reported from Canada and the USA from this source.

Paralytic Shellfish Poisoning (PSP) is caused by saxitoxin, a neurotoxin produced by some species of the dinoflagellates *Gonyaulax* and *Pyrodinium*. In blooms of these species, high concentrations of the toxin can be accumulated by bivalves, crabs, and some plankton-eating fish, generally without any harm to the organisms themselves, but these then present a hazard to those that eat them. Fatalities have been recorded in seabirds and fish at the time of these blooms. People who eat contaminated fish or bivalves are at risk of suffering paralytic shellfish poisoning with symptoms of nausea, loss of balance, defective vision, and, in severe cases, nausea, paralysis, and death. Areas regularly affected include the coasts of California, the Philippines, the Bay of Fundy, and north-east England, and shellfisheries have to be closed during the summer when the blooms are most likely to occur.

Neurotoxic Shellfish Poisoning (NSP) is similar to PSP but without the paralysis. The toxin is produced by the dinoflagellate *Gymnodinium breve* which forms blooms particularly off the coast of Florida.

OIL POLLUTION

Oil pollution of the sea attracts great public attention because it is visible and most people encounter it, either at first hand on bathing beaches or from pictures on television and in the press whenever there is a spectacular oil spill. Petroleum hydrocarbons reach the sea by many routes, however, and tanker accidents are by no means the only source of oil pollution.

INPUTS

It is difficult to calculate the total quantity of petroleum hydrocarbons entering the sea. Estimates vary between 1.7 and 8.8 million t year^{-1}. The best estimate, made in 1989, is about 2.5 million t year^{-1}, although some inputs have been reduced since then. The contributory sources are shown in Table 4.1.

Tanker operations

The world production of oil is about 3 billion t year^{-1} and half of it is transported by sea (Fig. 4.1). After a tanker has unloaded its cargo of oil it has to take on seawater as ballast for the return journey to the oil fields. The ballast water is usually stored in the cargo compartments which previously contained the oil. During unloading of the cargo, a certain amount of oil remains clinging to the walls of the compartment; this may amount to as much as 800 t in a 200 000 t tanker. The ballast water inevitably becomes contaminated with this 'clingage' which in any case has to be removed from the compartments before they can be filled with a fresh cargo of oil. Formerly, this

Table 4.1 Estimated world input of petroleum hydrocarbons to the sea (million t yr^{-1})

Source		Total
Transportation		
Tanker operations	0.158	
Tanker accidents	0.121	
Bilge and fuel oil	0.252	
Dry docking	0.004	
Non-tanker accidents	0.020	
		0.555
Fixed installations		
Coastal refineries	0.10	
Offshore production	0.05	
Marine terminals	0.03	
		0.180
Other sources		
Municipal wastes	0.70	
Industrial waste	0.20	
Urban run-off	0.12	
River run-off	0.04	
Atmospheric fall-out	0.30	
Ocean dumping	0.02	
		1.380
Natural inputs		0.250
Total		2.365
Biosynthesis of hydrocarbons		
Production by marine phytoplankton		26 000
Atmospheric fall-out		100–4000

dirty ballast and clingage were discharged to sea during the return voyage and were responsible for much of the casual oil pollution of the world's oceans. Measures have been introduced to reduce or eliminate this input.

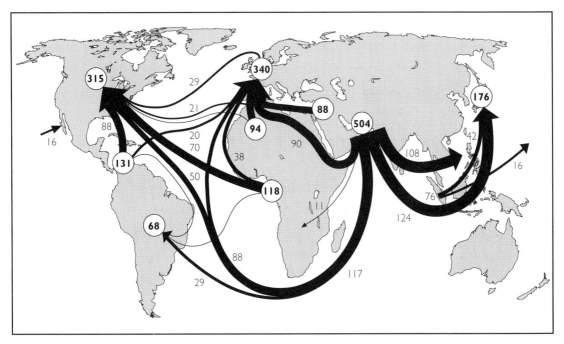

Fig. 4.1 Sea borne movements (in million t) of crude oil, 1989.

In the **load-on-top** system (Fig. 4.2), cargo compartments are cleaned by high pressure jets of water. The oily water is retained in the compartment until the oil floats to the top; the water underneath containing only a little oil is then discharged to sea and the oil is transferred to a slop tank. This process is continued until all the compartments have been cleaned and the tanker carries only clean ballast water. At the loading terminal, fresh cargo is loaded on top of the oil in the slop tank, hence the name of the technique.

The load-on-top method reduces, but does not totally eliminate, the discharge of oil into the sea. Increasingly, the clingage is removed by jets of crude oil (**crude oil washing**) while the cargo is being unloaded. To avoid the danger of explosion of the petroleum vapour generated during this process, the empty compartments have to be flooded with inert gas derived from the engine exhaust.

A further development has been the introduction of **segregated ballast** in which ballast water is carried in separate compartments and does not come in contact with the oil, so that

it can be discharged to sea without causing pollution. This is being introduced progressively as new tankers replace old ones, but in severe weather, even tankers equipped with segregated ballast may have to take on additional ballast water into the cargo tanks; this dirty ballast is subjected to the load-on-top process before it is discharged.

With the introduction of these improved methods of deballasting, the amount of oil entering the sea as a result of tanker operations has steadily reduced from an estimated 1 million t year^{-1} or more in the mid-1970s, to 700 000 t in 1981, and to 158 000 t in 1989.

Dry docking

All ships, including oil tankers, require periodic dry docking for servicing, repairs, cleaning the hull, and so on. It is essential that all oil is removed from the cargo compartments of tankers and from the empty fuel tanks of all shipping, to avoid the risk of explosion from petroleum gases. Generally, the journey from the unloading terminal to the ship repair yard is a relatively short one and the load-on-top

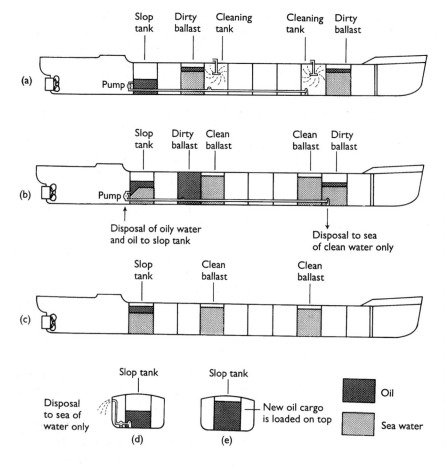

Fig. 4.2 Load-on-top. (*a*) Empty compartments are cleaned by water jet, (*b*) oil floats to the top in tanks containing dirty ballast, the water is discharged to sea and the oil transferred to the slop tank, (*c*) eventually the ship carries only clean ballast, oil floats to the top in the slop tank, and (*d*) underlying water is discharged to sea, (*e*) new cargo is loaded on top of the oil remaining in the slop tank.

system cannot be used effectively. Slop reception facilities have to be provided at the ship repair yards. Formerly, many yards did not provide such facilities, but the situation has now improved and the amount of oil reaching the sea in association with dry docking has been reduced from 30 000 t in 1981 to 4000 t in 1989.

Marine terminals

Accidents through human error and pipeline failure are an inevitable accompaniment to loading oil on to tankers and discharging it at oil terminals. The small size of this input reflects the care taken to reduce such accidents to a minimum.

Bilge and fuel oils

All shipping may need to take on ballast water when travelling unladen or in bad weather. Ballast tanks take up valuable cargo space and are limited in size, so additional ballast may be carried in empty fuel tanks. When the ballast water is pumped overboard it carries oil into the sea. In addition, all shipping needs to pump out bilge water which invariably contains oil from the ship's engines. This oil can be removed by separators, which is required in some sea areas under international conventions, although there are undoubtedly many illegal discharges of oily bilge water. Individually, the quantity of oil released may be small, but since all shipping contributes, the total amount of oil entering the sea is considerable.

Tanker accidents

As with other shipping, a large number of accidents involving oil tankers happen every year. Most result in no, or quite trivial, spillages of oil; the accident may not rupture the cargo compartments, a damaged vessel may be salvaged, or its cargo may be off-loaded into other tankers. Major shipping disasters involving tankers are a different matter. In March 1978, the *Amoco Cadiz*, carrying 223 000 t of Arabian and Kuwait crude oil, was wrecked on the coast of Brittany and lost its entire cargo, causing immense damage over a wide area. The tanker *Exxon Valdez* ran aground in Prince William Sound, Alaska, in March 1989 and spilled 37 000 t of crude oil, which came ashore on nearly 800 km of coast. On the other hand, in January 1993, the tanker *Braer* was wrecked on the coast of Shetland in hurricane-force winds and lost its entire cargo of 85 000 t crude oil which disappeared in the sea with almost no effect.

Table 4.2 lists the largest tanker accidents. Fortunately, such disasters are rare events. Shipping hazards are greatest close to land and in narrow straits such as the Straits of Malacca or the Straits of Dover, and near the entrances to ports where the density of shipping is high. It follows that most tanker accidents are close to shore and, if oil is spilled, coastal oil pollution almost invariably results.

Non-tanker accidents

When a ship is in an accident, its fuel oil may be lost to the sea. Some cargo ships, particularly bulk carriers, are now very large and carry as much fuel oil as a 1960 oil tanker carried crude oil, so this source of oil contamination is not negligible.

Offshore oil production

The oil extracted from the seabed invariably contains some water (**production water**), which must be extracted before the oil is transported to the refinery. This is done by oil separators on the platform and the oil concentration in the water that is discharged is usually less than 40 parts per million (ppm), but, in total, this amounts to a substantial quantity.

When an oil well is being drilled, **drilling muds** are pumped down the well. These maintain a head of pressure and prevent a blow-out when oil is struck, cool and lubricate the

drill bit, and carry the cuttings back to the surface (Fig. 4.3). The drill muds contain water or, if the nature of the strata demands it, 70–80 per cent oil; originally the oil was diesel, but this has been replaced by low toxicity oil. While there is some attempt to separate the **oil-based muds** from the drill cuttings before they are dumped on the seabed beneath the platforms, they are inevitably still heavily contaminated. About 90 per cent of the petroleum hydrocarbons entering the North Sea as a result of offshore oil extraction are from this source. There are some proposals that oil-contaminated cuttings should be brought ashore for disposal.

Blow-outs are the uncontrolled release of oil from the well; they are potentially very damaging because of the great quantities of oil that may be released before the blow-out is brought under control. Great precautions are taken to prevent them; nevertheless, accidents happen occasionally.

- In April 1977, a well in the Ekofisk field in the Norwegian sector of the North Sea blew out and discharged 20 000–30 000 t of oil.

- In June 1979, a well in the Ixtoc field off the Mexican coast blew out and was on fire; it took nine months to bring it under control and in that time it is estimated to have discharged 350 000 t of crude oil into the Gulf of Mexico.

- Oil platforms in the Nowruz oilfield in The Gulf were a casualty of the Iran–Iraq War in 1983 and released 3500–4000 t of oil per day for a prolonged period because hostilities in the area hindered remedial measures.

- This loss was dwarfed by the sabotage of the oil terminals at the end of the Gulf War in 1991, when perhaps 1 million t of crude oil were released into the sea.

Gas blow-outs, as in the North Sea Piper field in 1988, are less likely to result in oil pollution.

Atmosphere

The incomplete combustion of petrol or diesel in motor vehicles results in petroleum hydro-

Table 4.2 Large oil tanker spills

Date	Tanker	Location	Cause	Spillage (t)*
19 Jul. 79	*Atlantic Empress*	Off Tobago	Collision	287 000†
28 May 91	*ATB Summer*	700 n.m. off Angola	Fire/Explosion	260 000
06 Aug. 83	*Castillo de Bellver*	70 n.m. off Cape Town, South Africa	Hull failure	257 000†
16 Mar. 78	*Amoco Cadiz*	Brittany, France	Grounding	223 000
11 Mar. 91	*Haven*	Genoa, Italy	Fire/Explosion	144 000
10 Nov. 88	*Odyssey*	700 n.m. off Nova Scotia	Hull failure	132 000
18 Mar. 67	*Torrey Canyon*	Scilly Isles, U.K.	Grounding	119 000†
19 Dec. 72	*Sea Star*	Gulf of Oman	Collision	115 000†
12 May 76	*Urquiola*	La Coruna, Spain	Grounding	100 000†
23 Feb. 77	*Hawaiian Patriot*	320 n.m. west of Hawaii	Hull failure	95 000†
15 Nov. 79	*Independenta*	Istanbul, Turkey	Collision	95 000
29 Jan. 75	*Jakob Maersk*	Leixoes, Portugal	Grounding	88 000†
03 Jan. 93	*Braer*	Shetland Isles, U.K.	Grounding	85 000
19 Dec. 89	*Khark 5*	120 n.m. off Morocco, Atlantic	Hull failure	80 000†
03 Dec. 92	*Aegean Sea*	La Coruna, Spain	Grounding	73 000†
15 Feb. 96	*Sea Empress*	Milford Haven, U.K.	Grounding	72 000
16 Apr. 92	*Katina P*	Off Maputo, Mozambique	Hull failure	72 000
06 Dec. 85	*Nova*	75 n.m. off Kharg Island, Gulf	Collision	70 000
06 Dec. 60	*Sinclair Petrolore*	Off Brazil	Fire/Explosion	60 000†
01 May 75	*Epic Colocotronis*	60 n.m. northwest of Puerto Rico	Fire/Explosion	60 000†
31 Jan. 75	*Corinthos*	Marcus Hook, Philadelphia, U.S.A.	Collision	53 000†
07 Jan. 83	*Assimi*	60 n.m. off Muscat	Fire/Explosion	52 000†
09 Aug. 74	*Metula*	Magellan Strait, Chile	Grounding	50 000
31 Dec. 78	*Andros Patria*	Off Cape Finisterre, Spain	Hull failure	50 000
13 Jun. 68	*World Glory*	90 n.m. off Durban, South Africa	Hull failure	48 000
09 Dec. 83	*Pericles G.C.*	200 n.m. off Doha, Qatar	Fire/explosion	46 000†
13 Jan. 75	*British Ambassador*	Pacific Ocean	Hull failure	44 000
01 Jun. 70	*Ennerdale*	Off Port Victoria, Seychelles	Grounding	41 000
29 Feb. 68	*Mendoil II*	340 n.m. off Washington State U.S.A.	Collision	40 000†
27 Feb. 71	*Wafra*	Off Cape Agulhas, South Africa	Grounding	40 000
28 Dec. 80	*Juan A. Lavalleja*	Arzew, Algeria	Grounding	40 000
11 Jun. 72	*Trader*	Off southwest coast of Greece	Hull failure	37 000
24 Mar. 89	*Exxon Valdez*	Prince William Sound, Alaska	Grounding	37 000
21 Oct. 94	*Thanassis A*	200 n.m. off Manila, Philippines	Hull failure	37 000
01 Nov. 79	*Burmah Agate*	Galveston, Texas, U.S.A.	Collision	36 000†
09 Jun. 73	*Napier*	Off Guamblin Island, Chile	Grounding	35 000†
18 Jan. 77	*Irenes Challenge*	200 n.m. off Midway Island, Pacific	Hull failure	34 000
26 Dec. 70	*Chryssi*	250 n.m. southwest of Bermuda	Hull failure	33 000
06 Feb. 76	*Saint Peter*	Off Esmeraldas, Ecuador	Fire/Explosion	33 000†
28 Jan. 72	*Golden Drake*	1200 n.m. east of Bermuda	Fire/Explosion	32 000†
28 Apr. 79	*Gino*	40 n.m. off Brittany, France	Collision	32 000
16 Aug. 79	*Ionnis Angelicoussis*	Cabinda, Angola	Fire/Explosion	32 000†
05 Nov. 69	*Keo*	120 n.m. southeast of Nantucket, U.S.A.	Hull failure	31 000
25 May 76	*Caribbean Sea*	Off El Salvador	Hull failure	28 000

*Estimates of the size of spillages vary, sometimes widely
†Fire burned part of the spill

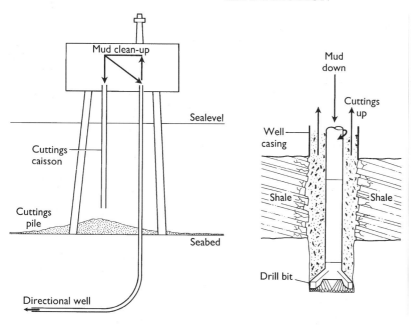

Fig. 4.3 Diagram of typical cutting discharges and drilling operations in the North Sea.

carbons being released into the atmosphere. Incomplete combustion of gas flared-off at oil platforms is another major source. These hydrocarbons are washed out in rain, either directly into the sea or indirectly by contributing to river runoff.

Municipal and industrial wastes

Domestic wastes and sewage contain a quantity of oils and greases and, depending on the nature of the industry, industrial wastes may also contain a considerable quantity of petroleum hydrocarbons. In some areas, these wastes are discharged into the sea and even if subject to treatment, may still retain petroleum hydrocarbons. This input is being steadily reduced.

Coastal oil refineries

Oil refineries are under severe pressure to minimize the oil content of waste water discharged to sea. Most achieve a standard of five ppm or less. Formerly, refineries used a steam cracking process and the waste water from them contained up to 100 ppm of oil. A few such refineries still exist but they are being phased out. The estimated input from this source given in Table 4.1 (see p. 38)

reflects the situation in 1989; present inputs are less.

Urban and river runoff

Every time it rains, irridescence caused by oil and petrol can be seen on the roads. This is washed down drains and into water courses, and eventually reaches the sea. Garage forecourts sustain a large amount of spilled oil which is washed into the drains.

Success in recovering used engine oil purchased for use in motor vehicles varies widely. In Germany and the Netherlands, virtually all such oil is recovered and refined or used as a fuel. In Britain, France, and Italy only 25–30 per cent of the engine oil sold is recovered. With motorists increasingly making their own oil changes, it must be suspected that much of this 'lost' oil finds its way into drains or on to the land, from which it is carried into rivers and then to the sea.

Licensed dumping at sea

Shipping channels in estuaries and ports commonly need regular dredging. The dredging spoil, which is usually dumped at sea, is contaminated with oil. Various kinds of solid municipal and industrial wastes that are dumped at sea may also contain petroleum hydrocarbons.

Natural sources

Oil deposits close to the earth's surface seep out and have done so for millennia. The pitch lakes of Trinidad are the product of natural seepage, and coastal oil seeps occur in many parts of the world.

Biosynthesis

Oil deposits are produced by plant remains that have become fossilized under marine conditions (coal results under freshwater conditions). Living plants produce hydrocarbons; the aromatic scent of pinewoods is caused by them. Estimates of the annual production of hydrocarbons by marine phytoplankton and by land plants from which the products are rained out into the sea are inevitably vague, but, in any case, the figures dwarf the input of fossil hydrocarbons by several orders of magnitude. Recent and fossil hydrocarbons have different constitutions and may well have different effects on marine ecosystems, but the dominating input of hydrocarbons from plants needs to be borne in mind when assessing the effect of petroleum hydrocarbons.

WHAT IS OIL?

Crude oil is a complex mixture of hydrocarbons with 4–26 or more carbon atoms in the molecule (Fig. 4.4). Arrangements include straight chains, branched chains, or cyclic chains, including aromatic compounds (with benzene rings). Some polycyclic aromatic hydrocarbons (PAH) are known to be potent carcinogens. Sulphur and vanadium compounds are also included in crude oil and nonhydrocarbons may represent up to 25 per cent of the oil. The exact composition of crude oil varies from one oilfield to another: much of the North Sea oil is light, with little sulphur and is low in tars and waxes; oil from the Beatrice field in the Moray Firth and that to the west of Shetland is a very heavy, waxy oil which needs to be heated in order to pump it through pipelines. The composition of the crude oil also varies during the life of a single oilfield.

Crude oil must be refined before it can be used. Refining is essentially a distillation process with different **fractions** or **cuts** taken at different boiling ranges (Table 4.3). Light gasoline is the basis for petrol used in motor vehicles; naphtha provides feedstock for the petrochemical industry; the residue is used as bunker fuel in ships and power stations; and the higher fractions are used as tars, and so on. Many of the commercial products are further refined, made into particular formulations, and receive additives of other materials to suit them for their various purposes.

All components of crude oil are degradable by bacteria, although at varying rates, and a variety of yeasts and fungi can also metabolize petroleum hydrocarbons. Small, straight-, and branched-chain compounds degrade most rapidly, cyclic compounds more slowly. High molecular weight compounds, the tars, degrade extremely slowly.

Table 4.3 Refinery 'cuts' of crude oil

	Boiling range (°C)	Molecular size (Number of carbon atoms)
Petroleum gases	30	3–4
Light gasoline, benzine	30–140	4–6
Naphtha	120–175	7–10
Kerosene	165–200	10–14
Gas oil (diesel)	175–365	15–20
Fuel oil and residues	350	20+

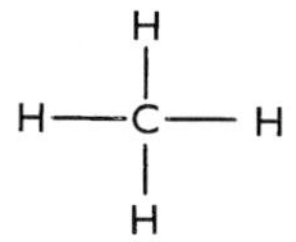

(a) Methane (CH$_4$) the simplest hydrocarbon

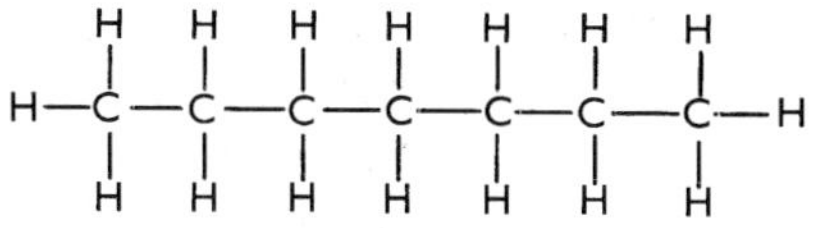

(b) A straight-chain alkane (or paraffin):heptane (C$_7$H$_{16}$)

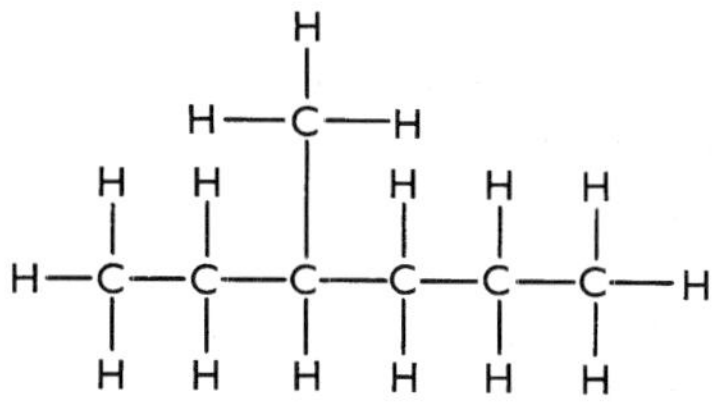

(c) A branched-chain alkane

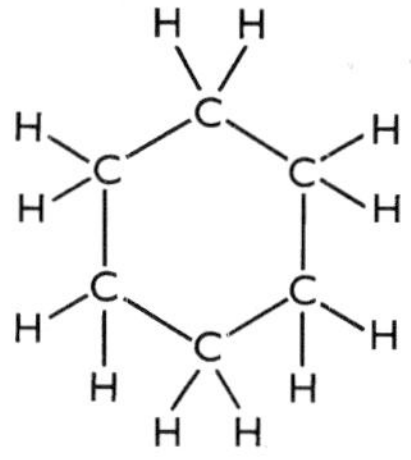

(d) A cyclo-alkane (naphthene)

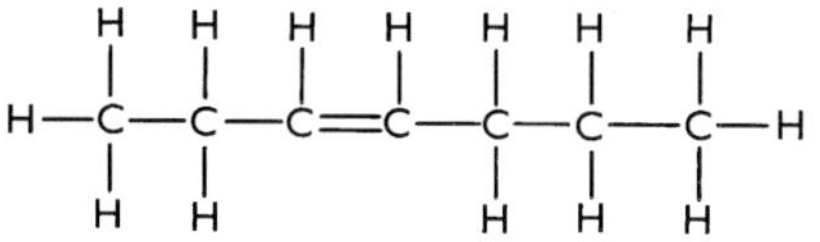

(e) An unsaturated hydrocarbon

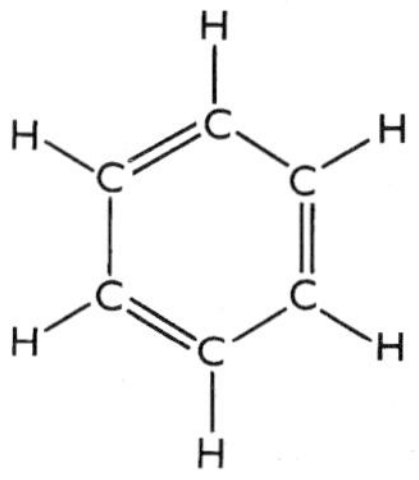

(f) Benzene, the simplest aromatic hydrocarbon

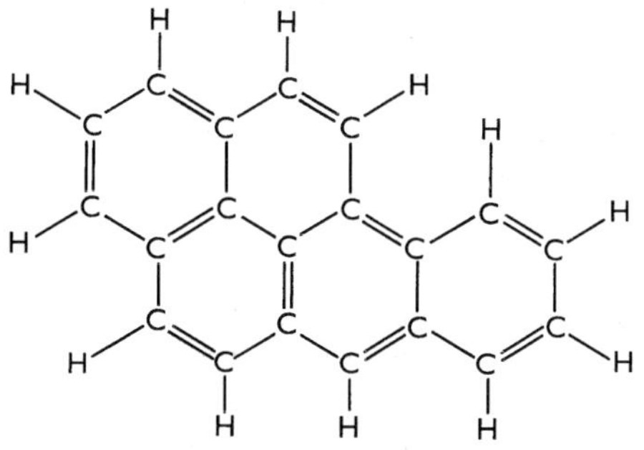

(g) Benzo [a] pyrene, a polycyclic aromatic hydrocarbon (PAH)

Fig. 4.4 The structure of some hydrocarbons.

FATE OF SPILLED OIL

When liquid oil is spilled on the sea it spreads over the surface of the water to form a thin film—an **oil slick**. The rate of spreading and thickness of the film depend on the sea temperature and nature of the oil; a light oil spreads faster and to a thinner film than a heavy waxy oil.

The composition of the oil changes from the time it is spilled (Fig. 4.5). Light (low molecular weight) fractions evaporate, water-soluble components dissolve in the water column, and immiscible components become emulsified and dispersed in the water column as small droplets. In some sea conditions a water-in-oil emulsion is produced; this may contain 70–80 per cent water and forms a viscid mass, known from its appearance as **chocolate mousse**. It forms thick pancakes on the water and intractable sticky masses if it comes ashore. The heavy residues of crude oil form **tar balls**, ranging in size from less than 1 mm to 10–20 cm in diameter.

Emulsified oil is in microscopic drops and therefore presents a large surface area at which bacterial attack can take place, so that the oil is rapidly degraded. In some circum-

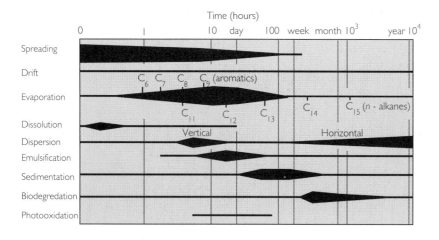

Fig. 4.5 The fate of oil spilled on the sea surface.

stances, the oil droplets may become attached to sediment particles in the water column and are carried to the seabed where the oil may be buried and bacterial degradation will be much slower. Tar balls and mousse present a small surface area compared with their volume and degrade extremely slowly for this reason. Tar balls of various sizes occur in the surface waters of all oceans, although they predominate in major shipping routes and in major ocean currents such as the Gulf Stream. They are mostly derived from tanker washings and routine shipping operations.

An oil slick does not remain in one place but travels downwind at 3–4 per cent of the wind speed (Fig. 4.6). In enclosed waters and estuaries, tides and water currents have a greater influence on the movement of an oil slick. If the oil becomes incorporated in the water column its movement is determined by water currents. This happened following the wreck of the tanker *Braer* on the south coast of Shetland in January 1993 and much of the oil was carried south-east in the Fair Island current, not, as forecast, blown north-east by the wind towards the coast of Norway (Fig. 4.7).

When a slick encounters land it is stranded on the shore with well-known consequences.

TREATMENT OF OIL AT SEA

Cleaning oil from beaches is difficult, time consuming, labour intensive, and costly. There

are obvious advantages if an oil slick that threatens the coast can be removed or otherwise dealt with while it is still at sea. A variety of techniques to do this have been developed, but only three have proved effective, and even these can rarely prevent beach pollution if a large amount of oil is spilled close to shore.

The natural process of emulsification of oil in the water can be speeded up by spraying chemical **dispersants** on the oil slick from ships or aircraft. Dispersants are not effective against heavy oils or oil that has been on the sea for some time (weathered oil). The first dispersants to come into use were highly toxic and there was some reluctance to use them, but low-toxicity dispersants have now replaced them. While dispersants are useful for treating small quantities of fresh oil, in any large spill their chief limitation is the difficulty of deploying sufficient ships or aircraft to spray more than a small proportion of the oil.

Seawater intakes, mariculture installations, or other important sites can be protected by the use of floating **booms**, with a 'sail' above the water-line and a 'skirt' below (Fig. 4.8), to deflect the floating oil to less sensitive areas. Large V-shaped booms may also be used to corral the oil, which can then be pumped out where it accumulates at the point of the V.

A variety of **slick-lickers** (Fig. 4.9) have been developed in which a continuous belt of absorbent material dips through the oil slick and is passed through rollers to extract the oil. Slick-lickers can deal with only small quanti-

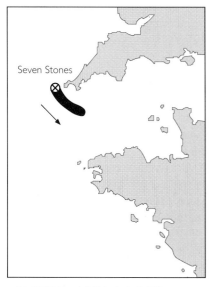

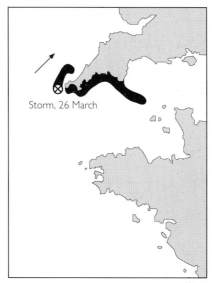

(a) 18-24 March, NW wind, oil drifting at sea

(b) 24-26 March, SW wind, oil ashore south Cornwall

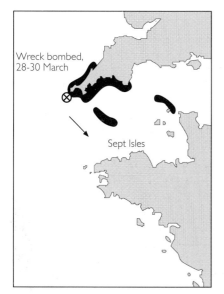

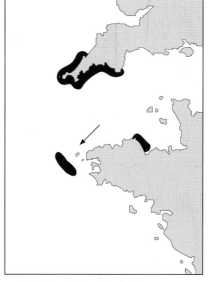

(c) 26 March-7 April, NW wind, oil ashore north Cornwall

(d) 8-12 April, NE wind, oil ashore north Brittany

Fig. 4.6 Movement of oil from the wrecked tanker *Torrey Canyon* in response to changing wind direction.

ties of oil and are useful in harbours and sheltered waters rather than the open sea.

BEACH CLEANING

When oil is stranded on the shoreline, biological damage may be caused by the toxic prop-erties of the oil if it is fresh, or by smothering if it is weathered. Once the detrimental effects of the oil have abated, natural recovery pro-cesses follow. Generally, clean-up of oiled beaches does not hasten recovery, although that is the intention, but often increases the damage and delays the recovery. Neverthe-less, much unnecessary and harmful oil spill

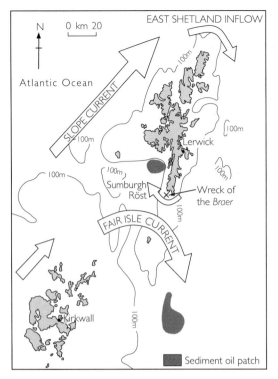

Fig. 4.7 Movement of oil from the wrecked tanker *Braer* in response to water currents.

clean-up takes place in response to public pressure that 'something must be done'.

A less aggressive approach to beach cleaning is **bioremediation**. This involves spraying

the stranded oil with nutrients, including phosphates and nitrates, to increase the rate of bacterial degradation of the oil. This technique was tested on a large scale in Alaska following the *Exxon Valdez* spill but it is not clear that it had much effect.

In some circumstances clean-up of the shoreline may be justified. If a large quantity of oil is stranded and the shore is accessible, much of the oil can be pumped into road tankers and removed, so reducing the chance of oil being washed off by a succeeding tide and redeposited elsewhere, thereby extending the area that is contaminated. A tourist beach may also have high priority for cleaning because the fauna and flora is of far less interest than the preservation of oil-free conditions.

The techniques of cleaning beaches depend on the nature of the shore.

• Rocks, harbour walls, and similar surfaces may be cleaned by high pressure water or steam, or by dispersants. If chemical agents are used they must be accompanied by a large volume of water in which the oil can be dispersed.

• On broken, rocky shores with boulders and stones of various sizes, low pressure water trickled across the beach for two or three days

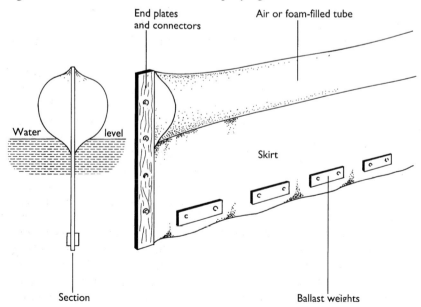

Fig. 4.8 Design of one type of floating boom.

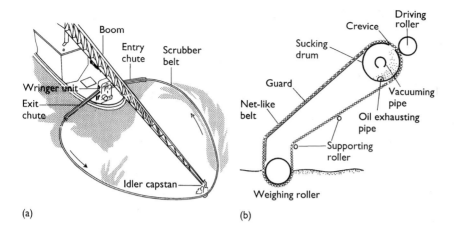

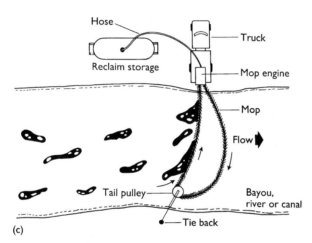

Fig. 4.9 Devices for mopping up floating oil—'slick lickers'. (*Elsevier Science*)

can be used to wash the oil to the water-line where it may be recovered or dispersed in the sea.

• Dispersants are useless on pebble or sand beaches because the dispersed oil merely drains into the beach; it disappears from the surface only to reappear at a later date. The only means of cleaning such beaches is to remove the surface layers of the substratum, either manually or by bulldozer.

• These are drastic treatments and the damaging effects on the fauna and flora can be reduced by using straw or cut vegetation as an absorbent to mop up much of the oil. On sheltered rocky shores with a good algal growth, a large amount of oil is trapped in the seaweeds, which can be cut and gathered.

Physical removal of oil by these methods results in only partial cleaning and much oil remains.

Unless beach oil can be dispersed into the sea, any beach cleaning operation produces a large volume of oil-contaminated debris. A few tonnes of oil can easily result in hundreds of tonnes of oily sand, pebbles, and other debris, and disposal of this material presents a serious problem. It cannot be incinerated unless the oil content is very high; it is usually unsuitable for dumping on waste tips because of the risk of the oil leaching out and contaminating water courses; and oil refineries cannot deal with it. Techniques exist for washing coarse sediments and it is possible to plough oily material into land where bacterial degradation eventually disposes of it, but neither technique is likely to be an effective way of dealing with large quantities of oiled material. At present, tipping on waste dumps where the oil can be contained appears to be the only practicable, if unsatisfactory, solution.

TOXICITY OF PETROLEUM HYDROCARBONS

The water-soluble components of crude oils and refined products include a variety of compounds that are toxic to a wide spectrum of marine plants and animals. Aromatic compounds are more toxic than aliphatics, and middle molecular weight constituents are more toxic than high molecular weight tars. Low molecular weight compounds are generally unimportant because they are volatile and rapidly lost to the atmosphere. A spillage of diesel fuel, with a high aromatic content, is therefore much more damaging than bunker fuel and weathered oil, which have a low aromatic content. A spillage of petrol or other 'white spirit' may present a serious fire hazard, but has little impact on marine organisms in the water.

ENVIRONMENTAL IMPACT OF OIL POLLUTION

Rocky shores

Rocky shores are high-energy beaches and stranded oil is quickly removed by wave action and water movement. The more sheltered the shore, the longer the oil remains and oil is removed most slowly from extreme high- and low-water levels and sheltered crannies where wave energy is least.

A considerable variety of animals and the more sensitive red and green algae are killed by exposure to fresh oil, but much of the oil reaching beaches is bunker fuel or crude oil that has been at sea for several days and lost most of its toxic constituents; it poisons few organisms. Limpets have been observed to graze on dried oil on the rocks without coming to any harm. But even if it is relatively

non-toxic, a weathered oil may still cause damage because of its physical properties.

Large amounts of stranded oil may kill animals by smothering them. On shores that received oil from the 1970 Santa Barbara blow-out in California, large *Balanus perforatus* projected above the oil and survived, but smaller species of barnacle were killed by smothering.

Many seaweeds secrete mucins which prevent oil adhering to them, but if it does adhere to the fronds storms may then tear the seaweeds from their stipes because of the increased weight of the fronds.

The rate of recovery and the form it takes depends upon the availability of colonizing forms. The experience of oil spills on rocky shores is that substantial recovery is usually achieved in two years but sometimes biological factors intervene and lead to prolonged change.

A well-known example of this is on some shores in Cornwall following the *Torrey Canyon* oil spill in 1967, where rocks previously covered with barnacles, *Balanus*, and limpets, *Patella*, became dominated by seaweeds, *Fucus*, and this change persisted for several years. This change was not caused by the stranded oil but by the very toxic oil spill dispersants then in use, which had, in the confusion, been poured neat over a few beaches. The dispersant, or the mixture of oil and dispersant, proved far more lethal than the oil alone and most organisms on the affected beaches were killed. Most critically, limpets were eliminated. In the absence of these dominant herbivores, diatoms and algae colonized the rocks and an algal succession continued unchecked during the summer following the oil spill. The presence of *Fucus* inhibited the settlement of *Patella* larvae when they were available later in the season, and also of barnacles. The change from a community dominated by *Patella* and *Balanus* appeared to be long-standing. Some four or five years later *Patella* became re-established and, surprisingly, were able to graze down the *Fucus* by rasping at the stipe which became weakened, and the fronds were then torn off. More than ten years after the spill, although barnacle domination of the rocky shore community had been restored, the associated fauna appeared to be less rich and varied than before. Cornish beaches not treated with dispersants were in an essentially normal condition again within two years of the spill.

A similar consequence of the loss of dominant herbivores has been reported in a sublittoral rocky community. In March 1957, a small tanker, the *Tampico Maru*, was wrecked and lost its entire cargo of diesel fuel at the mouth of a small bay on the Mexican coast of Baja California. The dominant sublittoral herbivores were the abalone *Haliotis* and two species of sea urchin, *Strongylocentrotus*, and these were either killed or left the affected area. A dense growth of giant kelp, *Macrocystis*, followed and persisted in the bay before the herbivores became re-established and brought the kelp under control (Fig. 4.10).

Soft substrata

Stranded oil is not readily removed from low-energy sedimentary beaches and, if still liquid, it drains down into the substratum. Here, the low oxygen concentration does not favour bacterial degradation of the oil, which may therefore retain its toxic properties for some time. When the *Amoco Cadiz* was wrecked on the Brittany coast in March 1978, releasing 223 000 t of crude oil, a considerable amount of it was carried into estuaries and inlets where it became incorporated in sediments. In addition to the immediate damage to the fauna, the persistence of toxicity prevented the start of recovery processes, and oil leaching from the sediments a year later caused renewed contamination.

In September 1969, the barge *Florida* ran aground during a storm in Buzzard's Bay, Massachusetts, near West Falmouth, and spilled 10 000 gallons (nearly 40 000 l) of diesel fuel. Although this was a relatively small oil spill, the toxic nature of the oil (containing 41 per cent aromatics) and the circumstances of the spillage made it unusually damaging. Winds and strong surf drove the oil into Wild Harbor, on to beaches, and churned up bottom sediments with which the oil became incorporated and was carried down to the bottom. Subsequent sediment transport extended the contaminated area. There was an immediate kill of fish, particularly in shallow creeks and bays which shelter juveniles of commercial species such as flounders and blue fish, and the adults of a variety of bait fish. Lobsters, crabs, shrimps, and bivalves were killed in large numbers. Scallops were particularly badly affected; oysters and soft-shell

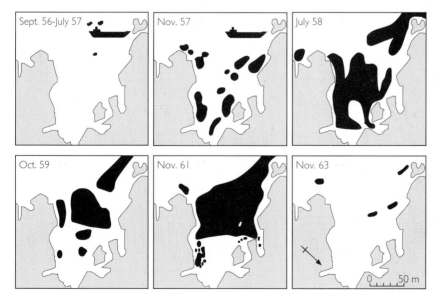

Fig. 4.10 Growth of giant kelp *Macrocystis* (shown in black) following the wreck of the tanker *Tampico Maru* on the Mexican Pacific coast.

clams less so. Commercial shellfish beds had to be closed because of tainting on a long-term basis owing to the continued risk of contamination by oil released from the shifting sediments. Detailed studies of the subtidal benthic community revealed instability persisting for more than five years in the most polluted parts of Wild Harbor; less contaminated areas began to show a successional recovery two years after the spillage.

Plankton

Plankton, and especially the neuston living in the top few centimetres of the sea, might be supposed to be particularly at risk because it is exposed to the highest concentration of water-soluble constituents leaching from floating oil.

Oil and oil fractions are toxic to a wide range of planktonic organisms, aromatic compounds more so than aliphatic ones. Weathered oil, after loss of volatile and water-soluble components, is not very toxic and, indeed, copepods have been observed to ingest oil particles and pass them through the gut without harm. Very low concentrations of petroleum hydrocarbons, below 50 ng g^{-1}, enhance photosynthesis, presumably because they have a nutritive effect. Above 50 ng g^{-1} there is a progressive depression of photosynthesis in algal cultures. At concentrations above 250 ng g^{-1}, feeding in the copepod *Acartia* is depressed and food selection is altered.

Studies of the response of whole plankton communities in the natural environment following oil spills have yielded conflicting results, but do not support the gloomy predictions from the laboratory studies. Following an oil spill from the tanker *Tsesis* in Swedish waters in 1977, zooplankton biomass fell dramatically for about five days by the death of the animals or their avoidance of the area. In the absence of the grazing zooplankton, there was a marked increase in phytoplankton biomass and productivity; it is not clear whether the oil had any nutritional effect. Following the wreck of the *Amoco Cadiz* in 1978, and the massive oil pollution of the Brittany coast, whatever initial impact there may have been on plankton, there was no sign of abnormality two to three weeks later, although there is a suggestion that the spring plankton outburst was depressed for a period.

Although it cannot be doubted that oil pollution, if sufficiently severe, kills planktonic organisms, it has not been possible to detect more than very transient effects, and sometimes not even those. Whatever losses are incurred are evidently rapidly made good, either by renewed growth or immigration from unaffected areas. Since petroleum hydrocarbons are subject to degradation by yeasts and bacteria, their main effect is to enhance primary production and act as a nutritive source.

Fixed vegetation

Salt marshes, sea grass beds, and, in the tropics, mangrove swamps are, like intertidal mudbanks, low-energy areas likely to trap oil, and the plants that form the basis for these ecosystems suffer accordingly. These are important ecosystems at the boundary between land and sea. They control coastal erosion, are a source of organic production which is transferred to the sea, and they provide shelter for the young stages of marine organisms, some of which are of commercial value.

The effect of oil pollution on annual plants living in a salt marsh depends on the season: if the plants are in bud, flowering is inhibited; if the flowers are oiled they rarely produce seeds; and if the seeds are oiled, germination is impaired. Generally, it may be expected that annuals will be killed by oiling and be dependent on reseeding from outside the area; recovery of annual populations may therefore require two or three seasons. Perennials show a range of reactions. Shallow-rooted plants with no, or small, food reserves, such as *Sueda maritima* or *Salicornia*, are readily killed. Others will generally survive at least a single exposure to oil, and perennials with large food reserves, such as those with tap-roots, survive repeated oilings. Generally the foliage is cut back, but decomposing oil has a nutrient effect and there is rapid and luxuriant renewed growth (a 'flush'). Experience of isolated oil spills suggests that oil pollution of this kind is less damaging to salt marshes than efforts to clean up the oil.

Mangroves present a rather different problem. They live in anoxic muds and have extensive air spaces carrying oxygen to the submerged part of the tree. Lenticels, by which the air is taken up, occur on the aerial roots of *Avicennia* or prop roots of *Rhizophora*, and, if the lenticels are clogged with oil, the oxygen level in the root air spaces falls to 1–2 per cent of normal within two days. Although mangroves have certainly suffered damage by oil spills, there are a number of cases where heavy oilings have not killed the plants. Too few studies of these tropical ecosystems have been made to allow a proper assessment to be made of their vulnerability to oil pollution.

Sea-birds

Whatever other effects oil pollution may have, the loss of sea-birds attracts the greatest public concern. It is difficult to give a precise estimate, but it is quite possible that tens or even hundreds of thousands of sea-birds are oiled in the north-east Atlantic every year. This sort of toll has continued ever since shipping converted from coal to oil-fired boilers, around the time of the First World War, and motor cars became numerous in the 1920s. It was feared that sea-birds might show a population decrease as a result of this heavy and persistent mortality.

Unlike most other organisms in the sea, sea-birds are harmed through the physical properties of floating oil, and the toxicity of its constituents is of minor importance. If liquid oil (or any other surface-active substance) contaminates a bird's plumage, its water-repellent properties are lost. If the bird remains on the sea, water penetrates the plumage and displaces the air trapped between the feathers and the skin. This air layer provides buoyancy and thermal insulation. With its loss, the plumage becomes waterlogged and the birds may sink and drown. Even if this does not happen, the loss

of thermal insulation results in rapid exhaustion of food reserves in an attempt to maintain body temperature, followed by hypothermia, and, commonly, death. Birds attempt to free their plumage of contaminating oil by preening and they swallow quantities of it. Depending on its toxicity, the oil may then cause intestinal disorders and renal or liver failure. Quite small quantities of oil ingested by birds during the breeding season depress egg-laying, and, of the eggs that are laid, the proportion that hatch successfully is reduced. If oil is transferred from the plumage of an incubating bird to the eggs, the embryos may be killed.

Indirect effects of oil pollution on reproduction appear to be much less important than the direct mortality of adult birds, and most attention has been directed towards the latter problem. The species most commonly affected are auks: guillemots (murres, *Uria aalge*), razorbill (*Alca torda*), and puffins (*Fratercula arctica*); and some diving sea ducks: scoters (*Melanitta nigra*), velvet scoters (*M. fusca*), long-tailed ducks (old squaw, *Clangula hyemalis*), and eiders (*Somateria mollissima*). These birds spend most of their time on the surface of the water and so are particularly likely to encounter floating oil, and because they dive rather than fly up when disturbed, they are as likely as not to resurface through the oil slick, so becoming completely coated with oil. Furthermore, these ducks are extremely gregarious except when ashore for breeding, and the auks are gregarious at all times of the year. Thus, if there are casualties they are likely to be numerous.

Indeed, quite small oil slicks drifting through concentrations of birds resting on the sea may inflict very heavy casualties quite disproportionate to the quantity of oil. When the 223 000 t of crude oil was lost from the *Amoco Cadiz* on the Brittany coast, the known sea-bird casualties numbered 4572. On the other hand, the much smaller oil spill of 37 000 t of crude oil following the grounding of the *Exxon Valdez* in Alaska in March 1989 resulted in over 30 000 known sea-bird deaths and probably very many more. Almost as large a death toll occurred in the Skagerrak in January 1981, when 30 000 oiled birds appeared on the beaches. This appears to have been caused by very small amounts of oil discharged by two vessels.

Other species besides auks and sea ducks are also regularly affected, although usually in much smaller numbers because of their habits and because they do not occur in such dense flocks. These include grebes (*Podiceps*) and divers (loons, *Gavia*), shags and cormorants (*Phalacrocorax aristotelis* and *P. carbo*), mergansers (*Mergus*), gannets (*Sula*), and pelicans (*Pelecanus*). The flightless penguins of the southern hemisphere occupy much the same ecological niche as guillemots in the northern hemisphere, but, except for the jackass penguin (*Spheniscus demersus*) around the coast of South Africa, the centre of population of most penguins is well to the south of heavily trafficked sea lanes or other sources of oil pollution, and so they rarely encounter oil slicks.

It is usually impossible to estimate more than crudely the number of birds killed by oil pollution. The only reliable figures are of the number of oiled birds or carcasses found on the shore; although an unknown proportion of the carcasses may have become oiled after death from other causes. An unknown number of oiled birds never reach the shore and, depending on wind speed and direction, sea conditions, distance from shore of the bird flocks, and accessibility of the coast to observers, it is likely that counts of oiled birds coming ashore underestimate actual casualties by anything up to 90 per cent. Figure 4.11 shows winter sea-bird casualties around North Sea and English Channel coasts, and the values reflect the interaction of these factors. The high proportion of birds stranded on the coast of the Netherlands that are found oiled, for example, reflects the prevalence of westerly winds in winter, causing drifting of both the bird flocks and any floating oil to the eastern side of the North Sea. The exceptionally heavy loss of sea-birds in the Skagerrak in January 1981 followed a period of westerlies which had funnelled large numbers of wintering birds into these narrow waters.

Auks not only include species that suffer

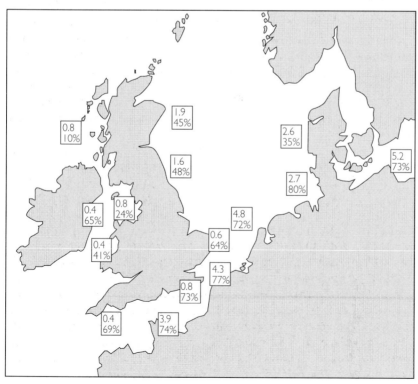

Fig. 4.11 Average number of dead sea-birds per kilometre found in beach surveys in late February, 1967–73, and the percentage of them that had been oiled.

most heavily from oil pollution, but they also have a reproductive pattern that does not favour rapid recovery from heavy adult losses. Guillemots, for example, do not start to breed until they are 3–4 years old, and thereafter do not breed every year; they produce only a single egg and the risk of losing eggs or chicks from predators or simply from rolling off precarious cliff ledges is such that only about 0.2 chicks per year are successfully reared for each breeding bird. The decline in guillemot and puffin colonies in south-western Britain, Brittany, and on the New England coast that had continued for much of this century, appeared to confirm that steady losses from oil pollution could not be made good by natural reproduction. Similar fears were expressed for the population of those species of sea duck particularly exposed to oil pollution, but since these produce numerous young there is less justification for concern.

Contrary to what had been predicted, while southerly colonies of puffins, guillemots, and razorbills were declining, more northerly colonies of these and other sea-birds have shown a dramatic increase in the last one or two decades, despite continuing oil pollution. The decline of southern colonies appears to have been the result of climatic changes in the north Atlantic in the period 1850–1950. The reasons for the growth of more northerly colonies is uncertain but appears to be a consequence of young birds starting to breed at an earlier age than before. It may be noted that occasional catastrophic mortality is not unusual for arctic and subarctic sea birds—a long period of severe weather which prevents feeding can cause as high a mortality as the worst oil spill—and despite their low reproductive potential, it would be surprising if a biological mechanism did not exist to maintain the population in the face of erratic losses of this kind.

Marine mammals

Although there have been occasional reports of seal pups being severely oiled, and possibly killed, by crude or bunker oil, there is no evidence that adult seals and sea lions or cetaceans are particularly at risk from oil on the sea.

Sea otters (*Enhydra lutris*) are exceptional among sea mammals. Unlike seals and cetaceans, which rely on subcutaneous blubber to provide thermal insulation, sea otters rely on their dense fur which functions in a similar way to the plumage of a sea-bird and they are similarly vulnerable to floating oil. Once driven to the verge of extinction by fur hunters, they are now common on the north-east Pacific coast, although rare elsewhere. A thousand or more were killed in the *Exxon Valdez* oil spill, in 1989, but with their high reproductive rate (populations are capable of growing at 17–20 per cent per annum until they reach a ceiling imposed by their food supply), the Alaskan populations are expected to recover quickly.

IMPACT OF OFFSHORE OPERATIONS

An offshore platform may suffer a blow-out, resulting in an uncontrolled discharge of oil, but fortunately this is a rare event. Routine inputs of oil to the sea are from production water (see p. 41) and small amounts of spilled oil washed off the platforms by rain. However, where oil-based drill muds (see p. 41) are used, bottom sediments receive a considerable amount of oil from contaminated cuttings.

The Ekofisk blow-out (see p. 41) had no detectable environmental impact, and even the very much larger Ixtoc blow-out (see p. 41) appears to have caused only minor damage on the beaches where oil came ashore.

No effect has been detected from oil discharged in production water or washed off platforms.

Drill cuttings dumped on the seabed, however, have a pronounced effect on the benthic fauna. The drill cuttings blanket the seabed, creating anoxic conditions and the production of toxic sulphides, with almost total elimination of the benthic fauna. Surrounding this area there is a recovery zone showing a succession of, first, opportunistic species such as *Capitella capitata*, then species able to tolerate stressful conditions and to out-compete the less tolerant species (Fig. 4.12), which gradually reappear further from the centre of disturbance as follows.

• An area up to 750 m from a production platform has hydrocarbon levels 1000 times the 'background' level and is characterized by a highly disrupted benthic community of low diversity.

• This is surrounded by a transition zone, between 750 and 1500 m, with hydrocarbon levels 20–100 times 'background'.

• The area between 1500 and 3000 m from the platform has a normal diversity of benthos (Fig. 4.13), but sensitive indicator species such as the polychaete *Aonides paucibranchiata* may be absent. Hydrocarbon levels are 5–20 times 'background'.

• There is no evidence of any effect on the benthos beyond 3000 m from the platform,

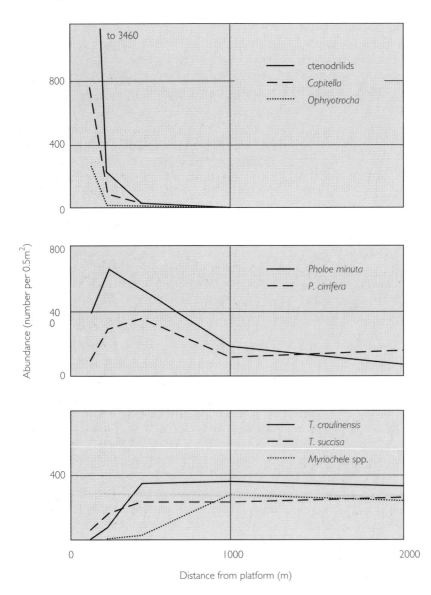

Fig. 4.12 Three types of abundance variation of benthic annelid worms influenced by disturbance at North Sea oil platforms.

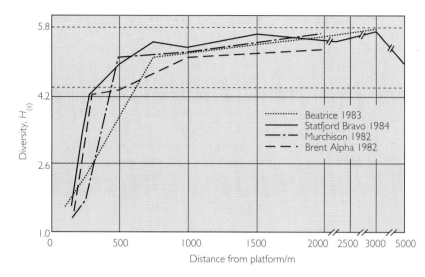

Fig. 4.13 Diversity of benthic fauna (measured by the Shannon–Wiener index) in relation to distance from North Sea production platforms. Broken lines indicate the maximum and minimum values obtained in studies made before oil operations began.

Legend within figure:
.... Beatrice 1983
———— Statfjord Bravo 1984
—·—·— Murchison 1982
— — — Brent Alpha 1982

Axis labels: Diversity, $H_{(s)}$ (vertical); Distance from platform/m (horizontal)

although elevated levels of hydrocarbons may be detected up to 10 km away.

PUBLIC HEALTH RISK FROM OIL POLLUTION

Some petroleum hydrocarbons are toxic to humans and there are a few cases on record of children being made seriously ill or even dying after inadvertently swallowing kerosene (paraffin). But humans have an extremely low taste threshold for petroleum hydrocarbons and the taste is particularly repulsive. There is therefore little risk of humans unknowingly receiving measurable doses of these toxins from contaminated food or drinking water.

Oil includes polycyclic aromatic hydrocarbons (PAH), some of which are known carcinogens. In the early 1970s there was a fear that PAH behaved in much the same way as chlorinated hydrocarbons like DDT—that they were resistant to bacterial attack and were excreted only slowly, if at all, by animals. As a result, it was concluded that these compounds might concentrate in the tissues of marine organisms with the concentrations increasing up the food chain to reach the highest levels in carnivorous fish. Human consumers of these fish might therefore be exposed to relatively large amounts of these carcinogens even in the absence of overt oil pollution. Fortunately, these fears have proved

to be groundless. There is little evidence that petroleum hydrocarbons accumulate in marine organisms, and seafood contains low concentrations of PAH (as measured by benzo [*a*] pyrene, a potent carcinogen) (Table 4.4) compared with those occurring naturally in some other foodstuffs that are eaten in far greater quantity (Table 4.5). Consumption of seafood is unlikely to contribute more than 2–3 per cent of the normal dietary intake of PAH.

COMMERCIAL DAMAGE FROM OIL POLLUTION

Fisheries

Fixed installations where fish or shellfish are held in intensive mariculture are particularly

Table 4.4 Concentration of benzo[*a*]pyrene (in $\mu g\ kg^{-1}$) in seafood from unpolluted and relatively polluted water

	Unpolluted	Polluted
Clam (*Mercenaria*)	0.38–1.1	8.2–16.0*
Shrimps	ND–0.5	ND–90.0
Crabs	ND–5.0	ND–30.0
Plaice	ND	0.05*
Herring	ND–0.1	0.4–13.0
Cod	ND–3.0	ND

ND = none detected.
*Evidence of tainting.

Table 4.5 Concentration of benzo[a]pyrene (in μg kg^{-1}) in various foodstuffs

	Benzo[a]pyrene content
Cooked meats and sausage	0.17–0.63
Cooked bacon	1.6–4.2
Smoked ham	0.02–14.6
Cooked fish	0.0
Smoked fish	0.3–60.0
Flour and bread	0.1–4.1
Vegetable oils and fats	0.4–36.0
Cabbage	12.8–24.5
Spinach	7.4

vulnerable to damage from accidental oil pollution because the animals cannot escape. A slick of oil drifting through such an installation may inflict commercial damage quite incommensurate with the size of the spillage. Japan, where mariculture is more varied and more advanced than in most other countries, has had many such unfortunate experiences.

In the open sea and, as a rule, in inshore waters, adult fish of commercial importance appear to be able to avoid areas affected by floating oil and are rarely killed. During the period when oil from a blow-out in the offshore oilfield at Santa Barbara, California, contaminated coastal waters, spotter planes used by the local fishing fleets to detect fish shoals, found them in normal abundance in areas of clear water between the streamers of oil.

Since fish eggs and larvae are more sensitive to toxins than adults, and are commonly in surface waters where they are likely to encounter high concentrations of petroleum hydrocarbons, they are expected to be particularly vulnerable to damage by oil pollution. In the laboratory, it has been found that Ekofisk crude oil reduced the hatching success of fertilized capelin (*Mallotus villosus*) eggs at concentrations of 10–25 nl l^{-1}. Concentrations of 250 μl l^{-1} cause developmental abnormalities in several species, which at best reduces their efficiency and in all probability results in their early death. Indeed, exposure of cod eggs to water-soluble extracts of Iranian crude

oil for 100 hours caused immediate casualties, and these continued even after the larvae were transferred to uncontaminated water.

Many commercial species of fish and shellfish produce enormous quantities of eggs, and even massive additional juvenile mortality may have no influence on the stock of adults that form the basis of the fishery.

Following the wreck of the *Torrey Canyon*, considerable areas of sea were sprayed with the very toxic dispersants then in use. A 90-per cent kill of pilchard eggs (*Sarda pilchardus*) was observed in the areas of spraying and a 50-per cent kill over a much wider area. Cornish pilchards form a small, isolated population but no shortfall could be detected two to three years later when the affected year-class were old enough to start appearing in the commercial catch.

One case is known in which an oil spill appears to have had a perceptible impact on fish stocks, and that was following the wreck of the *Amoco Cadiz* on the Brittany coast in 1977, and the devastating pollution that resulted. The one-year-class of flatfish is, in this instance, thought to have been reduced.

On theoretical grounds, it has been calculated that a major oil spill at the time, and in the place, of the main fish spawning in the North Sea would, at worst, cause a slight reduction in the overall catch for one year. In all probability, the effect would never be detected.

There have been reports of tumours and fin erosion in fish from areas chronically polluted by oil, and tumours and precancerous conditions in bivalves collected in such areas. Such pathological states occur naturally, but their incidence is increased in any polluted waters. They do not appear to be particularly related to petroleum hydrocarbons.

Most of the harmful effects of oil on fisheries refer to shellfisheries, either intertidally or in shallow water, and the damage may persist for years. A spillage of 700 t of diesel fuel (No. 2 fuel oil) in Buzzard's Bay, West Falmouth, Massachusetts in 1969, contaminated shellfish beds, salt marshes, and beaches, and the oil became incorporated in the sublittoral sediment. Some eight months after the accident, an area of 20 km^2 was polluted. Effects on crab populations were still obvious seven years later. In addition, following the grounding of the *Arrow* in Chedabucto Bay, Nova Scotia in 1970, 8000 t of Bunker C contaminated 240 km of shore-line. Populations of the clam *Mya arenaria* were still adversely affected six years later.

Not all shellfish beds suffer in this way. A sudden decline in production of the Louisiana oyster fishery in 1932–3 coincided with the beginning of coastal oil extraction and gave rise to fears that the subsequent intensive development of inshore and offshore oilfields would be inimical to the oyster industry. In fact, this did not happen The oyster fishery has had fluctuating success; occasional declines are not related to the activities of the oil industry but to other factors, such as incursions of freshwater into oyster lagoons. In fact, offshore oil platforms represent the only hard substrata in these waters and have attracted an exotic encrusting and sessile fauna; accompanying these are exotic fish species, and oil platforms have proven to be greatly favoured by sports fishermen in the area.

Tainting

Serious as the losses resulting from deaths of fish and shellfish are, the most important commercial damage is from tainting. Light oils and the middle boiling range of crude oil distillates are the most potent source of taint, but all crude oils, refined products, refinery effluents, wastes from petrochemical complexes, exhaust from outboard motors burning an oil–petrol mixture, and a host of other sources can impart an unpleasant flavour to fish and seafood which is detectable at extremely low levels of contamination. 'Oily' or 'petroleum' flavours are generally repulsive to humans and fish tainted in this way is unmarketable. Low levels of contamination may produce indefinite, but certainly detectable 'off flavours' which are at least as damaging to the market.

The concentrations of oil in water necessary to cause tainting in finfish vary widely with the oil and the fish: fatty fish like salmon contract a taint much more readily than non-fatty species. In the natural environment, finfish rarely become tainted from floating oil because they avoid the contaminated areas. The chief risk to commercial fishermen comes from fouling the fishing gear, which may result in tainting the whole catch. In chronically polluted waters, fish are either excluded or, generally not fished. Some chronically contaminated waters, such as Hong Kong harbour, are fished and the fish are tainted; the market in this case accepts tainted fish.

Shellfish may be affected in a variety of ways. Stranded oil contaminates the shells of molluscs without damaging them, but even a small quantity of oil on a few shells of mussels (*Mytilus*), cockles (*Cerastoderma*), or winkles (*Littorina*) can taint the whole catch if, as is commonly the case, they are boiled before eating. Since oil can be retained on the shells for a considerable time, this source of tainting is very persistent.

A commoner form of shellfish tainting that affects all filter feeders results from the uptake of fine droplets of emulsified oil in the course of normal feeding processes, and their subsequent incorporation into the tissues. Tainting in this way is common, and bivalves are the principal subjects. The oily taste is lost within three to four weeks if the source of contamination is removed or the shellfish are transferred to uncontaminated waters.

The fishing industry is vulnerable to irrational moves in a way that few other industries are, and even the suspicion that fish may be tainted or contaminated is sufficient to depress the market. When the wreck of the *Torrey Canyon* was in the news, fish sales on the Paris market fell by half, even though no fish from any area affected by the oil were ever on sale. There was a similar irrational response after the *Amoco Cadiz* wreck on the Brittany coast ten years later. Then, even cabbages grown in Brittany, far from the coast, were unmarketable for a period!

To avoid damaging their reputations in this way, and to limit the commercial losses, it is now usual to close fisheries in an affected area, as happened for the whole of the 1989 season for herring (*Clupea pallasii*) and pink salmon (*Oncorhynchus gorbuscha*) in Prince William Sound following the *Exxon Valdez* spill, and for farmed salmon (*Salmo salar*) and sea fisheries following the wreck of the *Braer* in Shetland.

Tourism

Tourists prefer their bathing beaches free from oil and most coastal resorts put a good deal of effort into removing tar and oily residues from their amenity beaches. This oil pollution, mostly resulting from casual discharges, is a nuisance but not necessarily a heavy financial burden on the resort because the beaches are cleared of litter, plastics, and other debris regularly in the course of preserving the local amenities, and removing odd patches of tar adds little to the cost of beach cleaning.

Severe pollution resulting from a major accident in the vicinity of a resort is a different matter. It may well be beyond the means of the local community to deal with it and is generally treated as a national emergency calling for a national response. Tanker wrecks, like those of the *Torrey Canyon*, *Betelgeuse*, *Arrow*, and *Amoco Cadiz*, are in this category.

It is commonly said that oil pollution of beaches is harmful to the tourist industry, but although tourists may grumble there is little evidence that they actually stay away from beaches and resorts subject to minor oil pollution. Even large-scale pollution, as from a tanker wreck, does not apparently deter tourists who have been known to flock to view a disaster or potential disaster.

METALS

CONSERVATIVE POLLUTANTS

Metals, like chlorinated hydrocarbons (discussed in Chapter 6), are **conservative** pollutants. Unlike organic wastes, which were considered in the previous chapters, conservative pollutants are not subject to bacterial attack, or, if they are, it is on such a long timescale that for practical purposes they are permanent additions to the marine environment.

Plants and animals vary widely in their ability to regulate their metal content; most can do so only over a limited range and metals that cannot be excreted remain in the body and are continually added to over the life of the organism. This is known as **bioaccumulation** and it dangers are illustrated by the following example. The pesticide DDT has about the same toxicity as aspirin. A lethal dose of aspirin is about 100 tablets; the same quantity of DDT is also lethal. But it is possible to take 0.5–1.0 g of aspirin a day indefinitely without ill effects because it is excreted. DDT is not excreted, so a lethal dose can be acquired after repeated exposure.

Animals feeding of bioaccumulators have a diet enriched in these conservative materials and if, as is commonly the case, they too are unable to excrete them, they in turn acquire an even greater body burden of the substance. This is **biomagnification**. Its chief significance is that top predators, including humans, may be exposed to very large concentrations of a conservative substance in their food. They are therefore a potential human health risk as well as a threat to natural resources, and they have been responsible for human deaths. For this reason, conservative pollutants are regarded with great concern.

Measures of contamination

Concentrations of contaminants in organisms are expressed as μg g^{-1} (parts per million, ppm) and μg kg^{-1} (parts per billion, ppb) (note that billion in this context is one thousand million).

The concentrations may be calculated on the basis of the **wet weight**, **fresh weight**, or **dry weight** of tissue. Wet weight is the weight of a sample of whole tissue removed from the body, fresh weight is the weight after being drained of free water, and dry weight is the weight after drying at 105°C to remove unbound water.

Tissues vary widely in their water content: feathers and hair contain practically no water, lobster shell is 25 per cent water, but the muscle tissue of scallops, *Pecten*, is 75 per cent water. Dry weight concentrations provide the best basis for comparing the concentrations of a substance in different tissues or different organisms, but many determinations are given only on a wet weight basis.

Concentrations of contaminants in sediments are always based on dry weight.

INPUT ROUTES

Assessing the effect of inputs of metals to the environment as a result of human activities is complicated by the very large natural inputs from the erosion of ore-bearing rocks, wind-blown dust, volcanic activity, and forest fires.

Table 5.1 World-wide emissions of trace metals to the atmosphere (in thousand t yr^{-1})

Metal	Natural sources	Anthropogenic sources
Arsenic	7.8	24
Cadmium	0.96	7.3
Copper	19	56
Nickel	26	47
Lead	19	449
Selenium	0.4	1.1
Zinc	4	314

Atmospheric inputs

An important route by which some metals enter the sea is by way of the atmosphere, to which there are large natural inputs of metals, such as aluminium in wind-blown dust derived from rocks and shales, and mercury from volcanic activity and degassing of the earth's crust. For some metals, such as lead, however, inputs to the atmosphere from human activities are greater, sometimes much greater, than natural inputs (Table 5.1).

Metals discharged to the atmosphere may exist as gases (mercury, selenium, and boron) or aerosols (most other metals). The length of time a contaminant remains in the atmosphere, and so the distance it travels in the air mass, depends on its reactivity, if it is a gas, or its density, if it is a particle. Lead has a residence time of about five days, gaseous organic compounds tens or hundreds of days. The metals are deposited by **gas exchange** at the sea surface, by fallout of particles (**dry deposition**), or are scavenged from the air column by precipitation (**wet deposition**).

It is difficult to estimate the atmospheric inputs of metals to the sea in different parts of the world and the figures given in Table 5.2 may be subject to considerable adjustment as more information becomes available.

Air–sea interactions are not a one-way process. Bubbles bursting at the sea surface release salt particles to the atmosphere and there is evidence that these particles become enriched with other contaminants during their formation. The sea is therefore a **source** of contaminants in the atmosphere as well as a **sink** for atmospheric contaminants. Too little is known about these processes to make very firm estimates of the relative importance of inputs to and outputs from the sea surface. Some idea of the possible scale of natural contributions from the sea to the atmosphere can be gained from the discovery that dimethyl sulphide synthesized by phytoplankton blooms of coccolithophores in the North Sea and released to the atmosphere may be responsible for half of the acid rain falling in coastal regions in summer.

River inputs

Most rivers make a major contribution of metals to the sea, the nature of the input depending on

Table 5.2 Transfer of metals from the atmosphere to the sea surface (in ng cm^{-2} yr^{-1})

Element	North Sea	Western Mediterranean	South Atlantic Bight	Tropical North Atlantic	Tropical North Pacific
Alumium	30 000	5000	2900	5000	1200
Manganese	920	—	60	70	9
Iron	25 500	5100	5900	3200	560
Nickel	260	—	390	20	—
Copper	1300	96	220	25	8.9
Zinc	8950	1080	750	130	67
Arsenic	280	54	45	—	—
Cadmium	43	13	9	5	0.35
Mercury	—	5	24	2.1	—
Lead	2650	1050	660	310	7.0

the occurrence of metal and ore-bearing deposits in the catchment area. Where a river passes through urban areas, the metal burden is augmented by human wastes and discharges.

Intense sedimentation in estuaries (p. 25) traps large quantities of metals which become adsorbed on to sediment particles and carried to the bottom. Sediments in industrialized estuaries with major ports contain the legacy of a century or more of waste discharge. Regular dredging of shipping channels in such areas produces large quantities of contaminated dredging spoil, which, except for the most heavily contaminated dredgings, are usually dumped at sea.

Other inputs

Much smaller quantities of metals are added to the sea by direct discharges of industrial and other wastes by pipeline, and by dumping sewage sludge and industrial wastes at sea. Although relatively small, these inputs may be locally significant if they are added to sea areas with restricted water circulation.

Table 5.3 shows the estimated inputs of some metals to the North Sea by various routes. It can be seen that rivers, the atmosphere, and dredgings are the most significant pathways; direct discharges and dumping at sea make substantially smaller contributions. However, the latter inputs are the easiest to control and are being further reduced.

UPTAKE OF METALS

Many metals are essential for living organisms. For example,

- the respiratory pigment haemoglobin, found in vertebrates and many invertebrates, contains iron;
- the respiratory pigment of many molluscs and higher crustaceans, haemocyanin, contains copper;
- the respiratory pigment of tunicates contains vanadium;
- many enzymes contain zinc;
- vitamin B_{12} contains cobalt.

Metals of biological concern may be divided into three groups:

- light metals (such as sodium, potassium, calcium), which are normally transported as mobile cations in aqueous solutions;
- transitional metals (such as iron, copper, cobalt, manganese), which are essential at low concentrations but may be toxic at high concentrations;
- metalloids (such as mercury, lead, tin, selenium, arsenic), which are generally not required for metabolic activity and are toxic to the cell at quite low concentrations.

Among polychaete worms, in particular, metals are accumulated for unexpected reasons. High concentrations of zinc in the jaws of nereids, and copper in the jaws of *Glycera* have a structural role in strengthening the tips of the jaws. *Nereis* also accumulates manganese in its jaws. Copper accumulated in the gills of the ampharetid *Melinna palmata* has the defensive function of reducing the palatability of the worm to its predators. The cirratulid *Tharyx marioni* has a high and constant level of arsenic in its feeding palps; its function is unknown.

Table 5.3 Estimated inputs (t) of some metals to the North Sea in 1990

Source	Mercury	Cadmium	Copper	Lead	Zinc
Rivers	25	43	1200	1000	6400
Atmosphere	6.9	74	740	1700	5500
Dredging spoil	19	71	1300	2700	7900
Direct discharges	1.8	17	290	160	1300
Industrial dumping	0.2	0.3	180	220	440
Sewage sludge	0.7	1.2	76	77	160

Absorption of heavy metals from solutions is dependent on active transport systems in some microorganisms and in sea urchin larvae, but, generally, in plants and animals it is by passive diffusion across gradients created by adsorption at the surface and binding by constituents of the surface cells and body fluids. An alternative and important pathway for animals is when metals are adsorbed on, or are present in, food, and by the collection of particulate or colloidal material by a food-collecting mechanism such as the bivalve gill. There is considerable variation in the extent to which plants and animals can regulate the concentration of metals in the body. Plants and bivalve molluscs are poor regulators of heavy metals, decapod crustaceans and fish are generally able to regulate essential metals such as zinc and copper, but non-essential metals such as mercury and cadmium are less well regulated.

MERCURY

Mercury is the only contaminant introduced by humans into the sea, apart from pathogens in sewage, that has certainly been responsible for human deaths.

Dissolved mercury in the sea is in the form of $HgCl_4^{2-}$ or $HgCl_3^{-}$, but, to a considerable extent, mercury is adsorbed on to particulate matter and is not in solution. It also forms stable complexes with organic compounds that occur in the sea, especially sulphur-containing proteins, or, in waters of reduced salinity, humic substances (see Table 2.2, p. 14). In anoxic substrata, mercury may be present as Hg, HgS, and HgS_2. Microbial systems in the sea convert all these inorganic forms of mercury into methyl mercury, which is readily released from sediment particles into the water and may then be accumulated by living organisms.

Sources and inputs to the sea

Natural inputs of mercury to the sea are from the weathering of mercury-bearing rocks and degassing of the earth's crust, particularly through volcanic activity. Forest fires and the biological formation of elemental mercury are additional natural sources. Estimates of natural inputs, especially that from degassing, vary widely. Formerly, they were thought to dwarf inputs resulting from human activities, but currently they are estimated to be of the same order, at 3600–4500 t amounting to 50–75 per cent of a total input of 6000–7500 t.

Following the discovery, in the early 1960s, of the dangers to human health of mercury in the marine environment, there has been a steady reduction of man-made inputs, partly through the imposition of strict controls on discharges of wastes containing mercury and partly through elimination of the use of mercury and mercurial compounds.

From a peak of 10 600 t in 1971, world production of mercury fell to a little over 3000 t in 1992. Table 5.4 shows the changing use of mercury in the USA, where the use of mercury in agricultural pesticides, pharmaceuticals, by the lumber and paper industries, and in antifouling paints has been progressively phased out. A similar pattern of decline in the use of mercury has been reflected in other industrialized countries, and Sweden plans to eliminate its use almost entirely by the year 2000.

These changes of practice have not been trouble free. Sweden banned the use of mercurial slimicides in the timber and paper pulp industry in 1967 and other countries followed suit. However, pentachlorophenol, which replaced the mercurial slimicides, proved to be damaging to marine macrofauna at concentrations of 75 $\mu g\ g^{-1}$ and accumulates in at least one marine organism, the polychaete *Lanice*. Pentachlorophenol then had to be withdrawn from use. The replacement of mercury by organic tin compounds in antifouling paints suffered similar problems (see p. 74).

Mercury inputs to the sea do not originate from industrialized countries alone. On the island of Palawan in the Philippines, cinnabar mine tailings were dumped in the sea at the rate of 100 000 t year^{-1} from 1955 to 1975 to form a peninsula 600 m long and 50 m wide.

Table 5.4 Consumption (t) of mercury in the United States

Use	1968	1974–75	1984	1988	1992
Electical apparatus	667	783	1170	207	124
Chlor-alkali industry	602	789	253	354	209
Paints	369	370	160	197	—
Industrial and control instruments	275	320	98	77	52
Dental	106	131	49	53	37
Agriculture	118	91	—	—	—
Catylists	66	82	112	—	—
Laboratory use	69	72	8	26	18
Pharmaceuticals	15	22	—	—	—
Paper and pulp	14	9	—	—	—
Amalgams	9	9	—	—	—
Other	298	206	48	55	145
Total	2628	2882	1798	1503	621

Cinnabar (HgS) has low solubility, but under oxic conditions it was converted to divalent and elemental mercury, most of which was then converted to the more toxic methyl mercury. Fish that are used as food by the local population were contaminated for several kilometres from the site.

Mercury used on a large scale to extract gold and silver in the Amazon basin results in a 100 t year^{-1} loss of mercury to the environment: 55 per cent to the atmosphere, 45 per cent to the river. Inhabitants of fishing villages downstream have up to 149.2 ppb in their blood, 99 per cent of it methyl mercury.

Although it has been possible to curtail direct inputs of mercury to the sea from industrial activities, it has proved more difficult to reduce discharges to the atmosphere. These are important because they are subsequently deposited by fallout over a wide area. Estimates of inputs to the atmosphere from the combustion of fossil fuel, municipal waste, sewage sludge, and smelting vary between 1000 and 6000 t year^{-1}. Systems to reduce emissions of particulates and sulphur dioxide in flue gases are not particularly effective at removing mercury.

Mercury in algae and invertebrates

Organic forms of mercury are more toxic than inorganic salts. Table 2.3 (see p. 14) shows the 18 h LC$_{50}$ of a number of mercury compounds to the red alga *Plumularia elegans*, and there is a similar pattern of toxicity of these compounds to larvae of the barnacle *Elminius* and the brine shrimp *Artemia*. As with other metals, bivalve molluscs take up mercury from the surrounding water very quickly. Figure 5.1 shows the uptake of mercury by the commercial oyster *Crassostrea virginica* exposed to mercuric acetate. Once the molluscs are returned to uncontaminated water, the mercury is lost from the body.

Mercury in fish

Most species of fish in oceanic waters contain 150 μg kg^{-1} (0.15 ppm) mercury in muscles, although much higher levels can be found in fish from contaminated waters. Large hake (*Merluccius*) from the Tyrrhenian Sea off the coast of Tuscany, where there is a natural input from mercury-bearing ores, may contain up to 3.2 ppm. Cod (*Gadus morhua*) taken from the Sound between Denmark and Sweden, which is heavily contaminated with mercury, contain 1.29 ppm; those from the North Sea, which is not seriously contaminated with mercury, contain 0.15–0.20 ppm, but those from Greenland contain only 0.01–0.04 ppm.

Some species, notably tunny (*Thunnus* spp.), swordfish (*Xiphias gladius*), and marlin

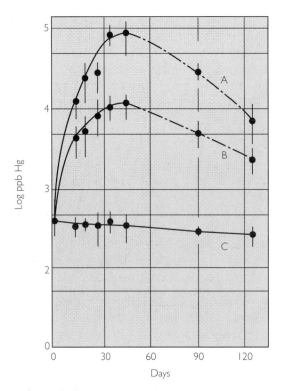

Fig. 5.1 Uptake of mercury by the oyster *Crassostrea virginica* exposed to (A) 100 ppb and (B) 10 ppb mercuric acetate for 45 days (——), and subsequent loss of mercury after transfer to uncontaminated water (– · – ·), (C) control. (*Springer-Verlag*)

(*Makaira indica*), naturally contain high concentrations of mercury. Concentrations of 1 ppm in the muscle are common, and may be as high as 4.9 ppm. Sharks (*Carchorhinus*) on the Florida coast contain up to 2.57 ppm mercury, and over 33 per cent of the fish exceed the recommended limit for human consumption of 1 μg g^{-1}. There are several reasons for this.

These fish are large carnivores at the end of food chains and their diets, therefore, contain high levels of mercury resulting from bioaccumulation and biomagnification. Food is not their only source of mercury; they are very active fish with a high metabolic rate and they swim continuously with their mouths open, so producing a forced flow of water across the gills. This results in a large uptake of oxygen, but also of metals (including mercury) dissolved in the water. Much of this mercury is in

the form of methyl mercury and since the fish cannot excrete it, its concentration increases with the age of the fish. Large, old specimens have high body burdens of mercury and warnings are given about the danger of eating the flesh of very large specimens of these species.

Halibut (*Hippoglossus hippoglossus*) is another long-lived species that accumulates high concentrations of mercury with age. A big halibut may weight 300 kg and be 50 years old. All specimens over 115 kg and half those weighting more than 60 kg may be expected to contain over 1 ppm of muscle mercury. As with most fish, over 90 per cent of this is in the form of methyl mercury and such fish are unsuitable for human consumption. They may be used for fish-meal, where the high concentration of mercury is diluted by the large mass of small fish containing little mercury that make up the bulk of the fish-meal used for animal feed.

Mercury in sea-birds

Sea-birds naturally acquire high concentrations of mercury without detrimental effect. High concentrations of mercury are found in the liver, where it may be demethylated, and selenium also accumulates in the tissues, where it gives protection against mercury toxicity.

The greatest concentrations recorded in sea-bird tissues are in wandering and sooty albatrosses (*Diomedea exulans* and *Phoebetria fusca*) from breeding colonies on Gough Island in the remote south Atlantic Ocean. They contained up to 271 μg g^{-1}* in feathers; such values would be harmful to land birds but there was no evidence of damage in these sea-birds.

A study of nestling ospreys (*Pandion haliaetus*) in different parts of Finland, some subject to mercury pollution, showed that there is a correlation between the concentration of mercury in the diet of the birds and that in the plumage; although in this species there is a limit to the amount of mercury that

*Because feathers contain practically no water, measurements are usually given on the basis of dry weight and therefore appear far higher than those for fish, which are on a wet or fresh weight basis (see p. 61).

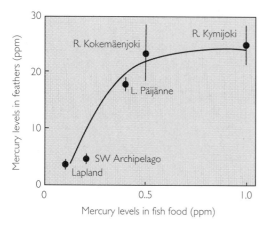

Fig. 5.2 Mercury concentration in feathers of nestling ospreys from five areas of Finland in relation to the mercury concentrations in local fish.

can be acquired by the feathers (Fig. 5.2), which is substantially below the high values recorded in the plumage of the albatrosses.

Most of the total body burden of mercury is stored in the plumage: about 80 per cent of the body burden in Cory's shearwater (*Calonectris diomedea*) and 93 per cent in Bonaparte's gull (*Larus philadelphia*) is in the feathers. Virtually all of it is methylated and it is shed when the birds moult. As much as 50 per cent of the remaining body burden of mercury is transferred to the growing feathers following the moult, so the plumage provides an important excretion pathway for methyl mercury.

The eggs of sea-birds are also tolerant to mercury contamination: reproductive dysfunction is observed in mallard (*Anas platyrhynchos*) with eggs containing 6–9 $\mu g\ g^{-1}$, but no effects on hatching and fledging occur in eggs of herring gull (*Larus argentatus*) contaminated with up to 16 $\mu g\ g^{-1}$ of mercury.

Mercury in the plumage of museum specimens of fish-eating birds reflects the rise and fall in mercury pollution over the last century. Feathers of white-tailed eagle (*Haliaetus albicilla*) in Swedish museums contain an average of 6.6 $\mu g\ g^{-1}$ mercury in specimens collected between 1880 and 1940, rising to about 50 $\mu g\ g^{-1}$ in 1964–5 specimens. Mercury contamination of the plumage of Baltic

guillemots (*Uria aalge*) rose by two to five times in the first half of this century (although it started to decline again after the mid-1960s), and feathers of Baltic guillemots contain three to five times as much mercury as those from Faeroese or Greenland birds.

Mercury in sea mammals

Marine mammals accumulate large quantities of mercury from their food, apparently without coming to harm. As in sea-birds, selenium antagonizes the toxic effects of mercury, and in seals, sealions, and dolphins, the selenium concentration in the tissues keeps pace with the amount of mercury. Mercury selenide found in the connective tissue of the liver of dolphins is apparently the product of a detoxifying mechanism for the methyl mercury acquired from their food.

Human health

The toxicity of mercury to humans has been known for centuries. Mad hatters in Victorian England got their name from the convulsions and loss of neuromuscular coordination symptomatic of chronic poisoning from the mercury used in the treatment of felt in hat manufacture. Inorganic mercury can be readily excreted, and, while it is dangerous to those exposed to it occupationally, it is not a hazard for the general public.

Organic mercury, such as methyl mercury, on the other hand, cannot be excreted easily and so may accumulate to toxic concentrations as a result of intermittent exposure over several years. It is a cumulative poison and since it can cross the barrier from blood vessels in the brain into the nervous tissue (the blood–brain barrier), it causes progressive and irreversible brain damage.

As far as is known, human exposure to methyl mercury occurs only through the consumption of contaminated fish and seafood. This was brought to light by an outbreak of methyl mercury poisoning in the small Japanese coastal town of Minamata (**Minamata disease**). Part of the population was dependent on fishing for a livelihood. The only industry was a factory that began producing vinyl chloride and acetaldehyde in 1952, both processes involving the use of mercury catalysts, large quantities of which were lost in washing the product and were discharged into the bay. The illness first appeared in 1953 and affected only fishermen and their families, but it was not diagnosed as metal poisoning derived from fish and seafood taken from Minamata Bay until 1956. In all, 2000 cases were recognized; of these, 43 died during the epidemic and over 700 of the survivors were left with severe permanent disabilities.

Fishing in part of the bay was banned at the start of 1957 and the epidemic was halted, but it was not until 1959 that it was shown that mercury was the toxic element involved, and 1960 that the source was the factory effluent, which was discharged directly into the bay. In addition to methyl mercury produced by methylating bacteria from the inorganic mercury included in the waste water from the factory, 5 per cent of the mercury in the discharge was in the form of methyl mercury. During the investigations in 1959, sediments near the outfall were found to contain 200 ppm mercury, bivalves in intertidal areas contained 10–39 ppm (dry weight), and fish in the bay contained 10–55 ppm (dry weight) mercury, most of it methylated.

Public health standards

Following the Minamata disaster there was a greater appreciation of the risk of mercury poisoning from eating contaminated seafood. The World Health Organization (WHO) recommended a maximum tolerable consumption of mercury in food of 0.2 mg of methyl mercury or 0.3 mg of total mercury per week.

Any standard must take into account how much fish is eaten, as well as the concentration of mercury in it. In Japan, for example, a far wider range and quantity of seafood is eaten than in Britain, where the average per capita consumption of fish is only 20 g per day. Most countries have set statutory limits between 0.5 and 1.0 ppm fresh weight of total mercury, often with the higher limit applying only to species such as shark, swordfish, and tuna which are known to accumulate high levels of mercury.

The reason for this exception is that a substantial proportion of the catch would be excluded from the market if the more stringent limit applied to it. In some countries (Canada, Sweden, South Australia) the more relaxed limit of 1.0 ppm for species that naturally have a high mercury content is coupled with advice to the public to limit its consumption of these species.

CADMIUM

Sources and inputs to the sea

Cadmium is widely distributed in the earth's crust, but is particularly associated with zinc and is produced commercially only as a by-product of zinc smelting. Cadmium has been used in quantity since about 1950 as stabilizers and pigments in plastics, in electroplating, and in solders and other alloys. Fears about the threat of cadmium to human health resulted in a reduction in its use for these purposes, but this has been largely offset by its increasing use in Ni–Cd batteries, and world production remains roughly constant at about $19\,500$ t year^{-1}.

The amounts of cadmium released to the environment cannot be quantified but are from a variety of often diffuse sources.

- Fumes, dust, and waste water from lead and zinc mining and refining, as well as from cadmium production.

- Rinsing water from electroplating contains 100–500 ppm cadmium.

- The iron, steel, and non-ferrous metal industries produce dust, fumes, waste water, and sludge containing cadmium.

- Zinc used in galvanized coatings of metals contains about 0.2 per cent cadmium as an impurity; it is estimated that all this cadmium is lost to the environment through corrosion in 4–12 years.

- The wear of automobile tyres, which contain 20–90 ppm cadmium as in impurity in the zinc oxide used as a curing accelerator.

• Phosphate rocks may contain 100 ppm cadmium, and phosphate fertilizers are a source of cadmium in the environment.

• Coal contains 0.25–5.0 ppm, and heating oils an average of 0.3 ppm cadmium, an unknown amount of which is discharged to the atmosphere.

• Sewage sludge contains up to 30 ppm cadmium.

The total input of cadmium to the world's oceans is estimated at nearly 8000 t year^{-1}, about half of which is the result of human activities, the rest is natural. Rivers and atmospheric inputs are probably equally important. About 2900 t year^{-1} of the cadmium is deposited in bottom sediments (most on the continental shelf), but it is difficult to account for the rest. Known fates of cadmium in the sea do not balance the marine budget and the cadmium content of the sea may be slowly increasing.

Cadmium in marine organisms

Cadmium is not an essential element for any organism although, for unknown reasons, it enhances phytoplankton photosynthesis and growth at concentrations up to 100 ppm. Because of its association with phosphates, cadmium is assumed to be taken up by phytoplankton but, except in a few instances, it does not appear to accumulate in the food chain.

The euphausid *Meganyctiphanes norvegica*, feeding on phytoplankton containing 2.1 ppm (dry weight) of cadmium, produces faecal pellets containing 9.6 ppm (dry weight), but has a whole-body concentration of only 0.7 ppm (dry weight). At higher levels in the food-web, fish and sea mammals have low concentrations of cadmium, at most a few ppm stored chiefly in the kidney, and are able to detoxify it by the production of a metallothionein.

Zooplankton in the surface layers of the ocean evidently have high cadmium levels because petrels and the hemipteran sea skater *Halobates*, both of which feed on surface zooplankton far from sources of contamination, tend to have high concentrations of cadmium.

Petrels may contain 49 ppm (dry weight) in the liver and 240 ppm (dry weight) in the kidney; *Halobates* normally contain 33 ppm, but may contain as much as 330 ppm (dry weight) cadmium.

Molluscs accumulate large concentrations of cadmium. This is particularly so in the bivalve Pectinidae: *Pecten novae-zeelandiae* has been found with 2000 ppm (dry weight) in the liver, but 1900 ppm (dry weight) has been found in the oceanic squid *Symplectoteuthis oualaniensis*, and oysters, limpets (*Patella vulgata*), and the dog whelk (*Nucella lapillus*) also acquire high concentrations of cadmium. *Nucella* shows clear evidence of accumulation of cadmium: in the Bristol Channel this dog whelk has been found to contain 38 ppm of cadmium, but the barnacles on which it feeds contain only 0.15 ppm.

Studies in the cadmium-contaminated Severn estuary and Bristol Channel in south-west Britain, where the input is from natural sources, and in the Sörfjord, a branch of the Hardanger Fjord in Norway, which receives smelter wastes, have failed to show any ecological effect beyond the contamination of some members of the local fauna.

Public health

Cadmium achieved notoriety in the aftermath of the Minamata disaster when it was claimed to be responsible for an outbreak of **itai itai** disease in a Japanese village on the Jintsu river. This painful disease affected the bones and joints of old women and resulted in a number of deaths. At the time it was attributed to contamination of rice by cadmium from the effluent from a zinc smelter, but now it appears more likely to have been associated with malnutrition and vitamin deficiency.

In fact, there are no well-authenticated reports of cadmium contamination of seafood causing lasting damage. High concentrations of cadmium (173 ppm) and zinc 57 600 ppm) in oysters from the Derwent estuary in Tasmania caused nausea and vomiting in people that consumed them, but there were no further effects.

Cadmium and its compounds are included in the 'black list' of substances, which also includes mercury, that, by international agreement, may not be discharged or dumped in the sea. Since cadmium compounds in the sea appear to be much less dangerous than was previously thought there have been several proposals to remove it from the black list. These have not been successful. There is evidence that inhalation of cadmium results in lung disease, including cancer, but this does not appear to be a hazard arising from marine pollution.

COPPER

Sources and inputs to the sea

The natural input of copper to the marine environment from erosion of mineralized rocks is estimated to be 325 000 t year^{-1}. Inputs from human activities are localized and vary widely in their nature. About 7.5 million t year^{-1} of copper are produced for use in electrical equipment, in alloys, as a chemical catalyst, in antifouling paint for ships' hulls, as an algicide, and as a wood preservative. Several of these uses inevitably result in copper being transferred to the environment. Urban sewage contains a substantial amount of copper and this is reflected in enhanced concentrations in sediments at sludge dumping grounds (see Figs 3.9(b) (p. 30) and 3.10(b) (p. 32)). The municipal waste from Los Angeles is estimated to contribute 510 t of copper to the sea annually. Runoff from mine tailings in tin and copper mining areas results in high copper and zinc concentrations in the River Fal in Cornwall and Rio Tinto in Spain. Antifouling paint releases all its copper to the sea. This is not a negligible amount: some antifouling paints contain 500 g l^{-1} of copper and it is estimated that 180 t year^{-1} of copper entered the coastal waters of California between Santa Barbara and San Diego from this source alone, until copper was replaced by tin compounds in antifouling paints.

Copper dissolved in seawater is chiefly in the form of $CuCO_3$ or, in water of reduced salinity, as $CuOH^+$. It also forms complexes with organic molecules. However, copper is one of the metals readily removed from solution by adsorption to particles and it is estimated that 83 per cent of copper in the sea is in this form.

Copper in marine organisms

Copper is an essential element for animals and the highest concentrations are found in decapod crustaceans, gastropods, and cephalopods, in which the respiratory pigment haemocyanin contains copper. Excess copper is usually stored in the liver; 4800 ppm copper has been detected in the liver of *Octopus vulgaris* and 2000 ppm in the hepatopancreas of a lobster, *Homarus gammarus*. Oysters may acquire very high concentrations of copper, stored mostly in the wandering leucocytes, and these blood cells may contain 20 000 ppm copper and 60 000 ppm zinc.

Although plankton, fish, and shellfish from areas known to be contaminated contain higher concentrations of copper than those from uncontaminated areas, copper does not generally accumulate in food chains. A predatory fish, the marlin (*Makaira indica*), at the top of a food chain accumulates mercury (see p. 66) but has low concentrations of copper: 0.3–1.2 ppm (average 0.4 ppm, wet weight) in muscles and 0.5–22.0 ppm (average 4.6 ppm, wet weight) in the liver.

Despite the existence of a number of detoxifying and storage systems for copper, it is the most toxic metal, after mercury and silver, to a wide spectrum of marine life, hence its value in antifouling preparations.

The very long-standing contamination of the estuary of the River Fal in south Cornwall, south-west England, has been investigated in detail, with interesting results. One branch of the estuary, Restronguet Creek (Fig. 5.3), is fed by the Carnon River which drains a formerly very productive tin mining area. Alluvial tin has been recovered from Restronguet Creek for several thousand years, since the Bronze Age, and the river has received wastes from deep mining for several hundred years. In the nineteenth century, arsenic refining and

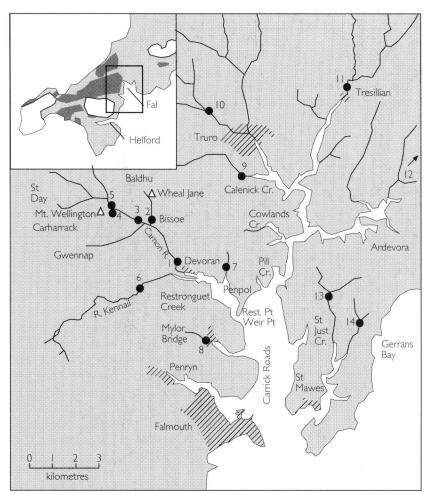

Fig. 5.3 River Fal and its tributaries. The Carnon River drains the chief tin mining area: 1–14 are sampling sites, △ indicates tin mines. Inset: west Cornwall showing metalliferous areas and the study area.

lead and tin smelting were carried out in the Carnon valley, adding to the contamination of Restronguet Creek. Although mining activity has now ceased, drainage from ancient mine tailings continues to carry a heavy load of metals into the creek. Levels of contamination of sediments are: zinc 2700 ppm, copper 2148 ppm, and arsenic 1732 ppm. These very high levels of contamination are about two orders of magnitude greater than for uncontaminated estuaries.

Despite this great contamination, the fauna and flora of Restronguet Creek is unexpectedly normal. Only bivalve molluscs, with the exception of *Scrobicularia plana*, and the estuarine gastropod *Hydrobia ulvae* are conspicuously absent. A variety of factors ex-

plains the presence of organisms in the water and sediment which might have been expected to be lethal.

Concentrations of copper in the polychaete *Nereis diversicolor* are closely related to those in the sediment, and worms containing more than 1000 ppm have been found in the creek. The copper is stored in the epidermal cells. Such levels would be lethal to *N. diversicolor* from uncontaminated estuaries (see Fig. 2.2, p. 12) and it appears that a copper-resistant strain of this species exists in the Fal. The alga *Fucus vesiculosus* in upper reaches of the creek accumulates several thousand ppm of copper and zinc (Fig. 5.4) and may also be a tolerant strain. Resistant strains are also known to occur in the fouling alga *Ectocarpus*

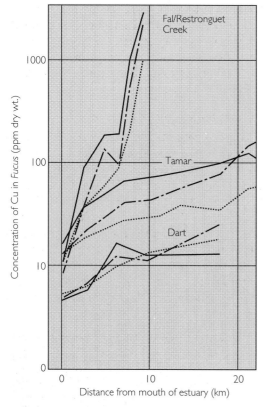

Fig. 5.4 Copper in brown seaweeds (*Fucus*) in three rivers in south-west England. Restronguet Creek receives mine waste, the Tamar passes through mining areas, the Dart is only slightly influenced by previous mining. · · · · growing tips; —— thallus; – · – · stipe. (*Elsevier Science*)

siliculosus which succeeds in growing on ships treated with copper-containing paint.

The polychaetes *Nephtys hombergi* and *Glycera tridactyla*, and the crab *Carcinus maenas*, also show increased tolerance to copper but are mobile species with distributed larvae, and they colonize the creek from less contaminated areas. There may be some selection of tolerant individuals, but copper tolerance may also be induced in these species through exposure to sublethal concentrations as they migrate into the creek.

Larval bivalves are very sensitive to copper and also to zinc, and it is not surprising that they are unable to survive in the high concentrations prevailing in the creek. *Scrobicularia*, which does occur there, is confined to the highest tidal levels where it is exposed to river water much-diluted by sea water. Breeding of this species is in July and August, and in dry autumn conditions with a low river input, juveniles have a chance to establish themselves.

Flounders (*Platichthys flesus*) feeding in the creek, largely on *Nereis diversicolor*, appear to be able to limit the assimilation of copper in the gut and do not acquire an increased body burden of the metal, although there is some evidence of liver damage in flounders exposed to high levels of metals.

Redshanks (*Tringa totanus*) wintering in the area, feed almost exclusively on *Nereis diversicolor* and the daily intake of worms by these birds is about equal to their body weight. They receive a high concentration of metals in their diet, but it is not known if they are affected by it.

The resilience of the Restronguet Creek fauna was shown in early 1992 when a large volume of untreated waste water that had backed up in a disused mine, burst out and inundated the creek. Although this water had a pH of 3.1 and very high metal concentrations (cadmium >600 μg l^{-1}) it had no significant effect on the local fauna.

Public health

Bivalves growing in contaminated water have the capacity to accumulate copper: the concentration factor for oysters is 7500 and they may accumulate so much copper that the flesh becomes green. This commonly happens to oysters in the River Fal and they have to be relaid in uncontaminated water for a year before they can be marketed. Humans are not at risk of copper poisoning from seafood: the lethal dose is about 100 mg, but the human taste threshold for copper is low (5.0–7.5 ppm) and the taste is repulsive. There is an ample safety margin and copper in the sea is not regarded as a health hazard.

LEAD

Sources and inputs to the sea

The total world production of lead is about 43 million t year^{-1}. Much lead in metallic form, in battery casings and plates, in sheet and pipes, and so on, is recovered and recycled, but most lead used in compound form is lost to the environment. Nearly 10 per cent of the world production of lead is used as petrol additives such as lead tetraethyl and it is lost, largely to the atmosphere. Globally, inputs to the atmosphere resulting from human activities, 450000 t year^{-1}, dwarf natural inputs of 25000 t year^{-1}.

Lead aerosols are carried to earth in rain and snow and are widely scattered. Figure 5.5 shows the lead content of the annual ice layers in Greenland from this source. The start of the industrial revolution marks the first major increase in the deposition rate; the growth in the number of automobiles using leaded petrol since 1940 is reflected in a second marked acceleration in deposition rates. The lead content of a peat profile in Derbyshire (Fig. 5.6) shows a similar increase.

Local, high concentrations of lead may be caused by special circumstances. Sea bass taken on the Californian coast near Los Angeles, with its very high density of automobiles, contain 22 ppm (wet weight) of lead in their livers. Corresponding figures for fish caught 300 miles offshore have 10 ppm, and off the Peruvian coast 9 ppm.

Sewage sludge dumping grounds may be expected to contain high lead concentrations.

Glasgow sewage sludge contains 771 ppm lead and sediments in the dumping area 200–320 ppm, whereas the general background in the Firth of Clyde is 48–139 ppm, and in an area far from sources of contamination such as Trinidad, is only 22 ppm.

Impact on marine organisms

Compared with other metals, lead in the sea is not particularly toxic and, at concentrations up to 0.8 ppm, lead nitrate even enhances the growth of the diatom *Phaeodactylum* (Fig. 5.7), presumably through the nutrient effect of the nitrate. Sublethal effects of low concentrations include a depression of the growth of *Cristigera* (a ciliate protozoan) by 8.5 per cent at 0.15 ppm and 11.8 per cent at 0.3 ppm (see Table 2.1, p. 12); the growth of the crustacean *Artemia* is significantly reduced at

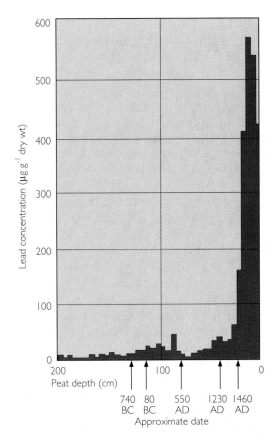

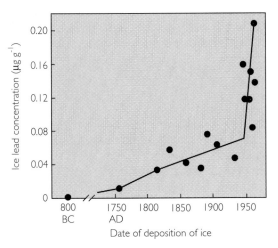

Fig. 5.5 Lead concentrations in annual ice layers in Greenland. (*Elsevier Science*)

Fig. 5.6 Lead concentrations in a peat profile in central England. (*Macmillan*)

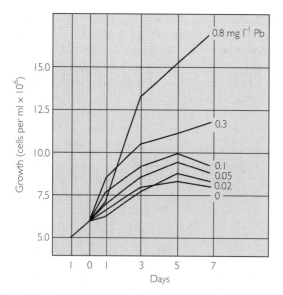

Fig. 5.7 Growth of culture of the diatom *Phaeodactylum* with lead nitrate in various concentrations added after the first day. (*Springer-Verlag*)

5–10 ppm; and the mortality rate of the mussel *Mytilus edulis* is increased by prolonged exposure to 10 ppm or less of lead salts.

Nevertheless, high concentrations of lead can be accumulated by some animals without apparent harm. The limpet *Acmaea digitalis* may contain about 100 ppm on the Californian coast near San Francisco where, evidently, the contamination is from aerial fallout. In the River Gannel in north Cornwall, where there is a natural input from lead deposits, estuarine sediments contain 2175 ppm lead, and the bivalve *Scrobicularia plana* living in them contains 991 ppm. In the contaminated Sörfjord in Norway, seaweeds and animals contain high levels, the content in mussels rising to 3000 ppm.

Mytilus has a detoxifying mechanism for lead and stores large quantities in the form of granules in the digestive gland. Because of their tendency to accumulate metals, mussels are regularly monitored in western European waters by the International Council for the Exploration of the Sea. Generally, the concentration of lead is below 1 ppm (wet weight), although occasional samples with up to 4 ppm have been recorded on the west coast of Sweden. Lead concentrations in mussels have declined in recent years in the North Sea and the Baltic, which probably reflects a reduction in the use of leaded petrol.

Fish contain little lead and the content in commercial species in the North Sea generally ranges from 0.05–0.15 ppm (wet weight). The livers of fish-eating seals and porpoises contain no detectable lead. Evidently, it does not present a hazard through bioaccumulation.

A rare instance of lead poisoning occurred among shore birds wintering on the Mersey estuary: 2400 birds, mainly dunlin (*Calidris alpina*), died in 1979 and a smaller number the following year. The dead birds contained more than 10 ppm (dry weight) of lead in their liver, 30–70 per cent of which was in the form of trialkyl lead. Further investigations showed that birds with 0.5 ppm trialkyl lead in the liver were at risk and some birds showed neuromuscular disorders symptomatic of lead poisoning. These waders acquired the lead from their food, which was predominantly the bivalve *Macoma baltica* containing 1 ppm lead, and the polychaete worm *Nereis diversicolor* containing 0.2 ppm. It is thought that a discharge of organic lead in factory effluent was responsible for contaminating the benthic fauna of the estuary.

Public health

Although lead is held to be responsible for serious damage to health on land, such contamination of the sea and marine products as occurs does not appear to be a matter for concern.

TIN

Antifouling paints containing organic tin compounds, tributyltin oxide (TBT) and tributyltin fluoride, are much more effective and long-lasting than mercury- or copper-based antifouling paints. They came into use in all classes of shipping in the 1970s and subsequently were also commonly used to treat net enclosures of mariculture installations and wooden lobster pots.

TBT is extremely toxic and is lethal to a variety of planktonic organisms, including mollusc larvae, which are 10–100 times more sensitive than the adults, at concentrations of only 1.0 μg l^{-1}. Such concentrations are commonly found in waters in and around yacht marinas and have been associated with recruitment failures of scallop and oysters in the affected areas.

The sublethal effects of low concentrations of TBT are of even greater consequence to commercial shellfisheries. Concentrations of TBT down to 0.01 μg l^{-1} cause reduced growth and other sublethal effects in very young Pacific oysters, *Crassostrea gigas*. More seriously, TBT causes gross thickening of the shells of oysters, greatly reducing the size of the animal inside, and rendering them unmarketable.

A curious and equally damaging sublethal effect of TBT has been discovered in intertidal, neogastropod dog whelks. A penis and sperm duct are developed in females, a condition known as 'imposex', evidently as a result of TBT triggering a change in the hormonal system. The oviduct and female genital opening are eventually overgrown by the developing penis and are blocked so that the egg capsules cannot be laid. Since dog whelks do not have a planktonic or other distributive stage in their life cycle, affected populations gradually decline and may be eliminated altogether because of this failure of reproduction. Most of the detailed studies have been made on the dog whelk *Nicella lapillus*, which has been seriously affected on north-west European rocky coasts exposed to inshore shipping traffic. It has now been found that this is a world-wide problem, and neogastropods (*Nucella*, *Thais*, *Cronia*, *Ilyanassa*, *Drupella*, *Naquetia*, and so on) are similarly affected in the north-western United States and Canada, Singapore, Malaysia, Indonesia, and New Zealand.

It is not known if TBT has similar damaging effects on other marine organisms, but that cannot be ruled out.

In mariculture installations where TBT is used to prevent fouling of nets and enclosures, organotin concentrations of 0.28–0.90 μg g^{-1} (wet weight) have been detected in the flesh of chinook salmon reared in them. This organotin is not destroyed by cooking.

Because of the damage to shellfisheries, environmental damage, and the possible threat to human health, the use of TBT in mariculture installations and on boats of less than 25 m length has been banned in the European Union, North America, Australia, and New Zealand. The use of TBT on ocean-going vessels is still permitted on the grounds that such large vessels do not congregate in inshore waters as small boats do, and the use of TBT antifouling paints on them is unlikely to result in damaging concentrations of TBT near shellfish beds.

TBT degrades in seawater to non-toxic compounds within a few weeks and the problems it has caused are unlikely to have a lasting effect. In areas where dog whelks have suffered a high incidence of imposex and a consequent reduction in population, there has been a good recovery within a few years of banning TBT antifouling paints. Surviving pockets of imposex may be owing to concentrations of large vessels at major ports, or, possibly, because imposex can be induced by agents other than TBT.

IRON

Iron is not usually a significant contaminant in the sea, but it came to prominence over sea dumping of red mud from the extraction of alumina from bauxite, and acid iron waste from the production of titanium dioxide; both wastes contain a large amount of iron. It was unfortunate that titanium dioxide manufacture created an environmental problem because it was introduced as a white pigment to replace lead oxides in paint, so reducing human exposure to lead.

Very large quantities of acid wastes were produced: western Europe produced 7.5 million t year^{-1} of acid iron waste, of which 5.6 were discharged to sea, either directly by dumping, or indirectly by pipelines or rivers.

Eastern Europe discharged about 2 million t year^{-1} to the Baltic and Black Seas and inland waters.

Because acid iron waste was discharged in such large quantities, concern was expressed about its possible environmental consequences by the fishing industry and conservation bodies.

It has proved very difficult to identify the environmental impact of acid iron waste discharges. The iron salts are precipitated as hydrated oxides of iron which drift around as flakes or particles before settling. A heavy deposit of iron oxides forms, like rust, on solid objects, including the shells and carapaces of animals. Particles adhere to fish eggs and larvae where they may clog delicate feeding structures. In one study, hydrated ferric oxide was found to cause loss of weight and increased mortality in mussels (*Mytilus edulis*), apparently through the increased production of pseudofaeces and loss of organic matter through the secretion of additional mucus. The oxides also appear to precipitate on the gills of fish, and other metals co-adsorbed with the iron are readily taken up.

It has proved impossible to reproduce in the laboratory the behaviour of iron oxides in the sea, but iron hydroxide in solution causes sperm agglutination, which results in reduced fertilization, and this could obviously have serious implications for animals that practise external fertilization. There is no evidence that acid iron waste discharges have affected local populations for this reason and field studies have yielded uncertain evidence about the impact of the iron in the waste. A survey at the dumping grounds in New York Bight in the early 1970s, after some 20 million t of titanium dioxide waste had been dumped over the previous 20 years, could detect no particular effects attributable to its iron content. On the other hand, this site had received a wide variety of toxic wastes for many years, it had an impoverished fauna and only hardy species remained.

Studies of plankton, benthos, and fish at a German dump site in the North Sea also failed to show any change, although iron concentrations in the area had increased. An attempt was made to relate the incidence of epidermal tumours and fin rot in fish to acid iron waste dumping in the area, but these conditions may arise in any area of waste disposal and appear to be a response to general stress and not to be related to a particular contaminant.

Investigations made around a sea outfall from a titanium dioxide factory in Natal showed that a cocktail of metals besides iron is included in the waste. As measured by the reduction in diversity and population size of the meiofauna, the increased concentrations of lead, vanadium, chromium, and zinc (in decreasing order of importance) all had a greater impact than iron (Fig. 5.8).

Subsequent to these investigations attention has focused on the acid sulphate content of the waste rather than the iron. Even so, despite the inconclusive evidence of harmful effects of discharges from titanium manufacture, there was sufficient concern that as a precaution, during the 1980s, sea disposal of titanium dioxide wastes was banned in the North Sea by the Netherlands and Germany, and the US Environmental Protection Agency recommended that the waste should no longer been dumped in New York Bight, but 150 km farther out to sea. Since 1995, the discharge of acid iron waste to sea by the titanium oxide industry has been greatly reduced; the acid sulphate is recovered as gypsum or sulphuric acid, while the iron and other metals are extracted and have found other outlets.

SOME OTHER METALS

Arsenic

Arsenic usually occurs as compounds with sulphur, either alone or in combination with metals. It has been used in pesticides and wood preservatives, in glass and enamel manufacture, and in alloys and electronics, but demand has fallen since the 1980s because of concerns about its possible environmental effects. Natural inputs to the sea are chiefly by rivers, particularly from areas of metalliferous mining, but some sea areas have historically received substantial inputs of arsenical indus-

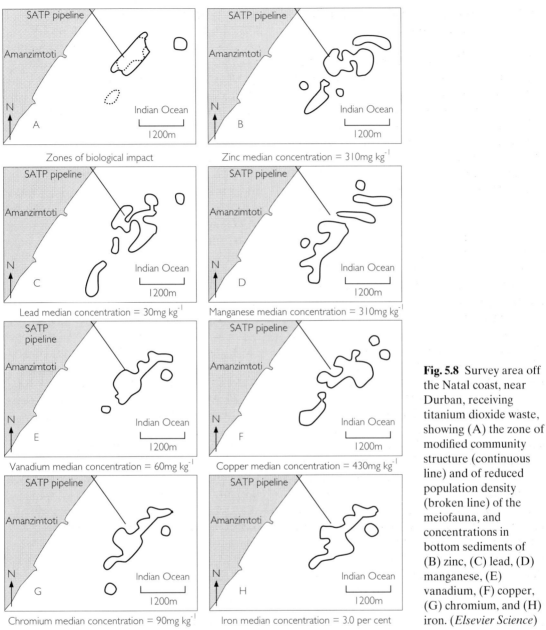

Fig. 5.8 Survey area off the Natal coast, near Durban, receiving titanium dioxide waste, showing (A) the zone of modified community structure (continuous line) and of reduced population density (broken line) of the meiofauna, and concentrations in bottom sediments of (B) zinc, (C) lead, (D) manganese, (E) vanadium, (F) copper, (G) chromium, and (H) iron. (*Elsevier Science*)

trial waste by direct outfalls and dumping.

The toxicity of arsenic depends very much upon the nature of the compound it forms and, particularly, its valency; trivalent arsenic is very much more toxic than pentavalent arsenic. Nearly all marine organisms contain arsenic. In marine animals it is chiefly in the form of arsenobutaine, which is pentavalent, very stable, metabolically inert, and non-toxic.

Marine algae do not contain arsenic as arsenobutaine but as carbohydrate compounds. A variety of aquatic organisms, including some algae, crustaceans, and fish, bioaccumulate arsenic, but it does not biomagnify through food chains.

Adverse effects on estuarine and marine organisms have been reported at arsenic levels of 100 μg l^{-1} and above. The 96 h LC$_{50}$

of arsenic trioxide for the zoaea larva of Dungeness crab (*Cancer magister*) is 232 μg l^{-1}, and the 48 h LC_{50} of sodium arsenite for abnormal development of Pacific oyster (*Crassostrea gigas*) is 326 μg l^{-1}. Field studies of the benthic fauna in areas where sediments are contaminated with arsenic have shown significant reductions in the abundance of polychaetes, molluscs, and crustaceans in Puget Sound at arsenic concentrations above 57 mg kg^{-1} (dry wt). Toxic effects have also been reported in marine benthos in Commencement Bay, Washington, and San Francisco Bay, California, at concentrations of 50–60 mg kg^{-1} (dry wt). All three areas are heavily contaminated with a variety of metals derived from many years of industrial discharges, however, and it is difficult to distinguish the effects of arsenic from those of other contaminants.

The toxicity of inorganic arsenic (arsenic trioxide) to humans is well known. The lethal dose is 50–300 mg, but varies widely between individuals. In several cases, inhalation of inorganic arsenic by workers in smelters has been associated with respiratory cancers, and chronic ingestion, as from drinking water, has been associated with bladder, liver, and kidney cancers, and possibly with skin cancer. The lack of toxicity of organic arsenic in seafood has been generally accepted by health authorities, partly because of reassuring toxicological data, but also because no poisoning episodes have ever been attributed to arsenic in the long history of human consumption of seafood. Less is known about arsenosugars present in seaweeds, but the use of seaweeds in east Asian countries, where they form a substantial part of the diet, suggests that there is no reason for toxicological concern. The Japanese hijaki (*Hizikia fusiforme*) is eaten in smaller quantities than some other species. Unusually, half its arsenic content is in the form of inorganic arsenate and heavy consumption of this product should probably be avoided.

Silver

Silver is used in photography, electric conductors, sterling ware, solders, coinage, electro-plating, catalysts, and batteries. It is a contaminant of many mining and smelting wastes and is associated with many sewage discharges. Geologically, silver is a rare element and natural concentrations in seawater are very low (0.1–0.3 ng l^{-1}), as they are in oxidized sediments (0.1 μg l^{-1}). Quite small inputs of silver from human activities can therefore increase local concentrations 200–300 times.

Silver can be bioaccumulated from solution by phytoplankton, some algae, and oysters. While low concentrations (0.1 ppm) are found in fish muscle, more than 1 ppm occurs in the liver. Similar concentrations have been found in the internal organs of shrimp, a cephalopod, a variety of gastropods, and the scallop *Pecten*. One of the highest values recorded was 68 ppm in the digestive gland of an oyster (*Ostrea sinuata*), but this is quite exceptional.

Silver in sediments is readily bioavailable and in one metal-contaminated estuary it was the most strongly accumulated of all the metals by the deposit-feeding bivalves *Scrobicularia plana* and *Macoma balthica*, and the polychaete *Nereis diversicolor*, which is also a deposit feeder. Although silver is bioaccumulated by a wide variety of animals, there is no evidence of biomagnification.

In bioassays, silver (with copper and mercury) is one of the three most toxic elements. The early life stages of bivalve molluscs are most sensitive and are affected by concentrations in the range <1–14 μg l^{-1}. However, natural concentrations of silver in seawater are several orders of magnitude less, and even in the polluted waters of San Francisco Bay, the concentration reaches only 0.025 μg l^{-1}, so there appears to be an ample safety margin.

Despite the great toxicity of silver, there is little good evidence that existing contamination levels are seriously damaging. In the following instances, silver has been held responsible for some environmental impact, although the possibility that contaminants other than silver are responsible or contributory factors cannot be ruled out: the absence of *Macoma balthica* from Fraser River sediments containing 2.1 ppm. silver; the low abundance of amphipods in southern Californian benthos (2.2 ppm); the low species richness in southern Californian sediments containing 2.5 ppm of silver; and increased burrowing time of *Macoma balthica* exposed to sediments from the Straits of Georgia, Washington (2.6 ppm).

Silver contamination of seafood has not been implicated in any health hazard to humans.

Nickel

Nickel is a heavy metal that is a significant contaminant of sediments in industrialized areas and serious attempts have been made to reduce inputs of nickel to the sea. It is used in steel and other alloys, electroplating, and batteries, and is also used as a catalyst. Fossil fuels are usually rich in nickel, and combustion of oil and coal results in a significant contribution to atmospheric deposition, but the major input of nickel to the sea is by rivers. Most of the nickel is particulate and there is heavy deposition in estuaries; dredgings from harbours and shipping channels are therefore often heavily contaminated. Municipal sewage sludge also contains significant quantities of nickel. Thus, input routes for nickel to the North Sea are estimated to be as follows: by rivers 2470 t year^{-1}, atmospheric deposition 1580 t year^{-1}, direct outfalls 650 t year^{-1}, dumping of sludge and other wastes 98 t year^{-1}.

Typical concentrations of dissolved nickel are: 0.2 μg l^{-1} (ppb) in the open ocean and 0.3 ppb in the open North Sea, but 1.0 ppb on the coast of the Netherlands. Concentrations of nickel in relatively uncontaminated sediments are 14 ppm at a site in Japan, 28 ppm in southwest England, and 30 ppm in west Scotland. Sediments in industrialized estuaries, of course, are much more heavily contaminated.

The toxicity of nickel varies widely and is influenced by salinity and the presence of other ions. Toxicity ranges reported from 96 h LC$_{50}$ tests include: phytoplankton 600 ppb (μg l^{-1}) to 1.0 ppm (mg l^{-1}), algae 2 ppm, polychaetes 17 to >50 ppm, crustaceans 150 ppb to 47 ppm, molluscs 60–320 ppm, estuarine fish 38–70 ppm, marine fish 8–350 ppm. The 48 h LC$_{50}$ for the larvae of the Pacific oyster (*Crassostrea virginica*) is 1.19 ppm and of the clam *Mercenaria mercenaria* is 310 ppb. Nickel is therefore regarded as only moderately toxic; the maximum acceptable toxicant concentration for freshwater fish is 380–730 ppb for nickel, compared with 62–125 ppb for lead or 9.5–40 for copper, and <0.26 for mercury.

No organisms have been found to contain very high concentrations of nickel. Bivalve molluscs commonly accumulate metals, but the highest concentrations of nickel that have been reported were in the scallop (*Pecten maximus*), with 22.9 ppm nickel in the kidney, 3.54 ppm in the digestive gland, and 0.04 ppm in muscle tissue. There is no evidence that nickel is bioaccumulated or biomagnified in marine food webs.

A survey made in the United States found that benthic species diversity is reduced in Massachusetts Bay where the mean nickel concentration in sediments is 21 ppm. More generally, the reduced diversity in Puget Sound, high oyster larva toxicity of Commencement Bay sediments, and the high toxicity of Los Angeles Harbour sediments are all associated with nickel concentrations of about 30 ppm. However, all these sea areas are contaminated with a variety of metals and other pollutants, and it is impossible to identify the particular contribution of nickel.

6

HALOGENATED HYDROCARBONS

Hydrocarbons containing chlorine, fluorine, bromine, or iodine (the halogens) differ from petroleum hydrocarbons (see Chapter 4) because most of them are not readily degraded by chemical oxidation or bacterial action. Like metals (see Chapter 5), they are **conservative pollutants** and are essentially permanent additions to the marine environment. Unlike metals, most of these compounds are man-made and do not occur naturally, and they tend to accumulate in both animals and sediments. The great majority of them contain chlorine and are known collectively as chlorinated hydrocarbons or organochlorines.

LOW MOLECULAR WEIGHT COMPOUNDS

Halogenated hydrocarbons embrace a very wide range of compounds. Low molecular weight hydrocarbons, particularly methane, are synthesized by marine algae and possibly by a few invertebrates, and may contain chlorine, bromine, or, occasionally, iodine. Elevated concentrations of these substances may therefore come from natural sources and not be the result of human activity.

Even so, low molecular weight volatile, halogenated hydrocarbons have been manufactured in very large quantities and almost all this production has been lost to the environment. They include the industrial solvents dichlorethane (CH_3CHCl_2) and vinyl chloride ($H_2C=CCl_2$), the solvents carbon tetrachloride (CCl_4), and perchlorethylene ($Cl_2C=CCl_2$) used in dry cleaning, and trichlorethane (CH_3CCl_3) and

trichlorethylene ($ClHC=CCl_2$). Another class of low molecular weight halogenated hydrocarbons are freons or carbofluorocarbons (CFCs) such as CCl_3F and CCl_2F_2. They are extremely stable, not flammable or toxic, and cheap to produce. They were originally developed as coolants for refrigerators and air conditioners, but later were also used on a very large scale as aerosol propellants and in foamed plastics.

Some low molecular weight chlorinated hydrocarbons, particularly the CFCs, are responsible for depletion of the ozone layer in the upper atmosphere. World production of CFCs reached more than 1 million t year^{-1} by 1986, but in the next year international agreement was reached in the Montreal Protocol to halve the use of CFCs by 1996 and phase them out altogether by the year 2000. In fact, world production of CFCs fell by 40 per cent between 1988 and 1992, and the European Union and USA ceased production and use of them by the end of 1995.

Volatile low molecular weight chlorinated hydrocarbons are not regarded as a particularly serious threat in the sea, but most, including carbon tetrachloride, chloroform, trichlorethane, and trichlorethylene, are, like CFCs, in the process of being phased out.

PESTICIDES AND PCBS

The higher molecular weight chlorinated hydrocarbons have been a matter of particular concern because, unlike the low molecular weight compounds, they do enter marine

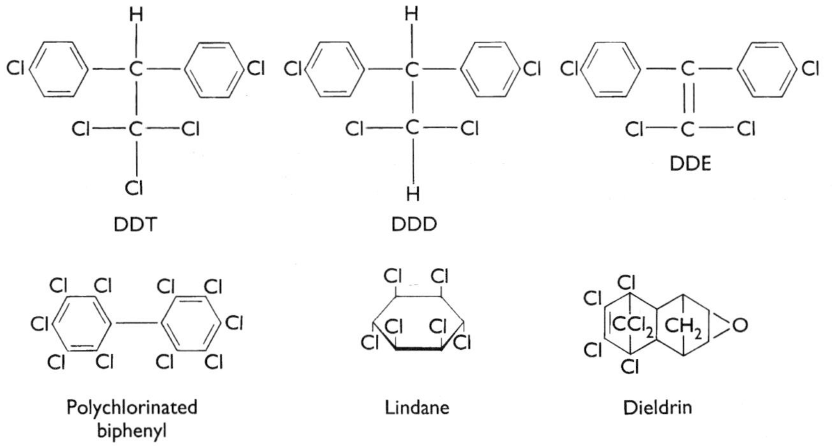

Fig. 6.1 The structure of some chlorinated hydrocarbons.

ecosystems and they accumulate in animal tissues, particularly in fatty tissues. These chlorinated hydrocarbons include several classes of pesticides, and the polychlorinated biphenyls (PCBs) (Fig. 6.1).

DDT and related compounds

Although DDT (dichlorodiphenyltrichloroethane) was known in the last century, it was not introduced as an insecticide until 1939. In many ways it is the ideal pesticide.

- It is extremely toxic to insects, but very much less toxic to other animals.
- It is very persistent, with a half-life in the soil of about 10 years.
- It continues to exert its insecticidal properties for a very long time.
- It is relatively cheap.

Few other pesticides share the desirable properties of being reasonably specific to the target organisms, of remaining effective for a long time so that only a single application may be needed, and of being safe for the humans exposed to it.

The use of DDT prevented epidemics of typhus among refugees and in war-damaged areas in Europe during and immediately following the Second World War. It provided the basis for the very successful programmes of the World Health Organization to control malaria and other insect-borne diseases in Africa and the Far East. Its primary use, how-

ever, particularly in developed countries, was pest control in forestry and in most forms of agriculture.

DDE (dichlorodiphenylethane) is a derivative of DDT, resulting from the loss of one chlorine atom from the —CCl$_3$ group in the DDT molecule. It has low toxicity to insects and is not used as a pesticide. Most of the chlorinated hydrocarbon in the sea, and 80 per cent of that in marine organisms, is in the form of DDE and presumably nearly all of it has been derived from the breakdown of DDT.

DDD (dichlorodiphenyldichloroethane) is another derivative of DDT resulting from the loss of a chlorine atom from the —CCl$_3$ group in the DDT molecule. It has some toxicity to insects and is less toxic than DDT to fish. For this reason it is occasionally used as an insecticide in situations where its low toxicity to fish is important. It can be excreted by many organisms and rarely accumulates in them.

Commercial DDT is a mixture of DDT, DDE, and DDD, with DDT predominating.

DDT has now been almost completely superseded by less persistent insecticides in the developed world, but it continues to be extensively used in developing countries because it is effective, safe to handle, and cheap.

'Drins'

This group of interrelated insecticides includes aldrin, dieldrin, endrin, heptachlor, and so on. They are all extremely persistent and the degradation products are also toxic: heptachlor degrades to heptachlor epoxide which is even more toxic than the parent compound; aldrin degrades to dieldrin.

These pesticides were used as seed dressings and in situations where their high toxicity and persistence were required, as in the control of wireworms (*Tipula* larvae) in ploughed up grassland. Their chief disadvantage is that they are toxic to mammals as well as being persistent and subject to bioaccumulation. They were largely withdrawn during the 1970s, although they continued to be used to mothproof textiles, carpets, and fleeces, and for some minor agricultural purposes, until the late 1980s.

Although now used only in very small quantities because of their persistence, 'drins' are widespread in the environment and continue to leach out of agricultural land into watercourses and the sea.

Lindane (γ-HCH)

Lindane (gamma-hexachlorocyclohexane, or γ-HCH) was previously, and sometimes still is, incorrectly known as benzene hexachloride (BHC) and this name may even be erroneously applied to hexachlorobenzene (see below).

Lindane came into use at about the same time as DDT and acts as a contact poison to insects. It is volatile and stable at high temperatures and so can be used as a smoke to fumigate crops in, for example, orchards. It is also used as a seed dressing and in the treatment of timber. It tends to accumulate in food chains and some attempt has been made to restrict its use; γ-HCH levels in cod livers in the southern North Sea fell from about 300 μg kg^{-1} in 1977 to about 50 μg kg^{-1} in 1987. However, it is still widely employed in agriculture and horticulture, particularly in India and China.

Hexachlorobenzene (HCB)

This product was formerly widely used as a soil fumigant and as a seed dressing for grain or as a fumigant in grain storage against fungal attacks. It is also used in wood preser-

vatives. It has largely been replaced as a soil fumigant, but continues to be used in agriculture for other purposes. It is used as a fluxing agent in aluminium smelting and also occurs as a by-product in the manufacture of carbon tetrachloride, pentachlorophenol, and vinyl chloride monomer. There are, therefore, numerous routes by which it may reach the sea. It is relatively insoluble in water and most of the HCB in the sea is attached to sediment particles. It is highly persistent.

Toxaphene

Toxaphene is the trade name for polychlorinated camphenes and was introduced as an insecticide in the mid-1940s. Its method of synthesis results in the production of a complex mixture of at least 670 camphene compounds with six to ten chlorine atoms per molecule and a total chlorine content of about 68 per cent by weight. It is used as an insecticide on cotton and vegetable crops, and in livestock dips. It is acutely toxic to fish and is occasionally used as a piscicide in freshwater lakes. Since its introduction it has been the most heavily used pesticide in the United States, with a total production not far short of that of DDT at its peak. Its use is now coming under regulation.

Polychlorinated biphenyls (PCBs)

PCBs have been in use since the early 1930s. They are not pesticides, but have been used in electrical equipment, in the manufacture of paints, plastics, adhesives, coating compounds, and pressure-sensitive copying paper. Because they are chemically very stable, they resist chemical attack and act as flame retardants. They have been used in fluid drive systems and as dielectrics in transformers and large capacitors.

The number of chlorine atoms per molecule varies from one to ten, but commercial PCBs are produced to physical, not chemical, specifications and contain a mixture of isomers. The product may contain 20 per cent by weight of chlorine and average about one chlorine atom per molecule, or up to 60 per cent chlorine with different percentages of biphenyls containing three to six atoms of chlorine per molecule.

Following concern about the environmental impact of chlorinated hydrocarbon pesticides, a number of steps were taken to reduce the use of PCBs as well. In 1970, Monsanto, the sole manufacturer of PCBs in the United States, voluntarily reduced production from 33 000 t in 1970 to 18 000 t in 1971. A variety of measures were subsequently adopted to limit the use of PCBs to functions where recovery and recycling is possible: dielectrics in transformers and large capacitors, and in hydraulic systems. By the mid-1980s most members of the European Union had stopped the production of PCBs except for purposes where there was the possibility of recovering and recycling the product. It is now planned that all PCBs will have been collected and safely destroyed by 1999, and considerable progress has been made towards that end. However, up to the time when the use of PCBs was restricted, total world production had amounted to more than 1 million t, most of which has been dispersed in the environment where it continues to give trouble.

Dioxins and furans

There are 75 different isomers of chlorinated dioxins containing one to eight chlorine atoms. The isomer TCDD (2,3,7,8-tetrachlorodibenzo-dioxin) has caused greatest concern. Chlorinated dibenzofurans are structurally similar to dioxins, but there are 135 isomers.

Dioxins are physically and chemically stable and tend to be increasingly stable with increasing halogen content. They are insoluble in water, but soluble in organic solvents, fats, and oils. They are extremely toxic and strenuous efforts are underway to eliminate them. The chief sources of dioxins were associated with

- the manufacture and use of chlorophenols in the wood processing and treatment industries;
- the manufacture and use of the herbicides 2,4,5-T and 2,4-D, which are a significant source of di-, tri-, and tetra-CDDs.

These and other industrial sources of dioxins have now been substantially reduced or eliminated.

The most important remaining source of dioxins is from the combustion of organic material, including fossil fuels, wood, and, particularly, the incineration of municipal and medical waste, and from domestic heating. The synthesis of dioxins is greatest at temperatures between 200 and 400 °C and emissions can be reduced by ensuring that when waste is incinerated, it spends the minimum time in this temperature range.

The European Union programme to reduce dioxin emissions from all sources by 70 per cent between 1985 and 1995 appears to have been largely successful.

INPUTS TO THE MARINE ENVIRONMENT

Aerial transport

The uses for which pesticides and most PCBs are designed results in them being widely distributed in the natural environment, but the principal source of widespread pesticide contamination is from the agricultural use of these compounds. Aerial transport is the main route by which they reach the sea (Table 6.1).

Low molecular weight chlorinated hydrocarbons are volatile. All organochlorine pesticides volatilize and, particularly in the tropics where they are still used in large quantities, climatic conditions favour their release to the atmosphere. Both γ-HCH and HCB are rapidly lost to the atmosphere in the presence of water vapour: in one study made in India, 99.6 per cent of the γ-HCH applied to rice paddies was found to be lost to the air. In Nigeria, 98 per cent of the DDT applied to a cow pea crop volatilized within four years.

Further to this, DDT, the 'drins', and toxaphenes adsorb strongly on to particles and are carried into the sea in wind-borne dust.

Some agricultural practices particularly favour the aerial distribution of pesticides.

- *Aerial spraying.* In the fruit and vegetable growing areas of California, where there is

Table 6.1 Comparison of atmospheric and river inputs of organochlorines to the world oceans (t yr^{-1})

Compound	Atmosphere	Rivers	% Atmospheric
ΣHCH	4754	40–80	99
HCB	77.1	4	95
Dieldrin	42.9	4	91
ΣDDT	165	4	98
Chlordane	22.1	4	85
ΣPCB	239	40–80	80

heavy use of pesticides, it has been estimated that 50 per cent of the pesticide from crop-spraying aircraft never reaches the ground but forms aerosols and, as such, may travel great distances.

• *Cultivation of arid areas*. Arid areas may be intensively cultivated with the aid of irrigation, but dry soil with its burden of adsorbed pesticides is transported in dust storms.

Aerial transport has resulted in the world-wide distribution of organochlorine pesticides (Table 6.2). DDT is an entirely man-made substance which does not occur naturally; it has been in use only since 1940. Yet, within 20 years, DDT and its residues had pervaded the entire biosphere; even king penguins in Antarctica, several thousand kilometres from any place where DDT had been used, contained detectable traces of it.

River inputs

Although the total burden of pesticides carried in to the sea by rivers is small compared

with aerial inputs (Table 6.1), it may be locally damaging. In addition to rain washing pesticides from the plants and soil into rivers and hence to the sea, irrigation water, particularly if it is applied by spraying, has the same effect. Floods carry very large quantities of silt into the sea and if the silt is derived from agricultural land, it may carry a considerable burden of adsorbed pesticide.

The use of DDT and the 'drins' was phased out in western European states in the early 1970s, but elevated levels of chlorinated hydrocarbons are still recorded near the mouths of the major rivers, and over 3 t year^{-1} of PCBs enter the North Sea from river inputs, mainly from the Rhine. It is possible that there is still some illegal use of organochlorine pesticides in the catchments of these rivers, although that is unlikely; most probably the input is from sediments still contaminated with pesticides carried in the runoff from the land, or of PCBs in drainage water from poorly maintained land disposal sites.

Table 6.2 Input of some organochlorines to the oceans (t yr^{-1})

Compound	North Atlantic	South Atlantic	North Pacific	South Pacific	Indian Ocean	Total
ΣHCH	851	97	2640	471	698	4757.0
HCB	16.8	10	19.9	18.9	11.4	77.0
Dieldrin	16.6	2.0	8.9	9.5	6.0	43.0
ΣDDT	15.6	14.0	66.4	25.7	43.3	165.0
Chlordane	8.7	1.0	8.3	1.9	2.4	22.3
ΣPCB	99.7	13.8	35.5	29.1	52.1	230.2

Direct inputs

Although direct inputs of chlorinated hydrocarbons to the sea have largely ceased, a large quantity of pesticides and PCBs from these sources continue to contaminate bottom sediments.

In the manufacture of chlorinated hydrocarbons, a variety of related compounds are usually produced, only one of which is the desired product. During purification, these unwanted hydrocarbons (generally unknown) are extracted and were formerly dumped at sea in considerable quantities.

Chlorinated hydrocarbons often appeared in industrial outfalls to the sea. One striking case is that of the Montrose Chemical Company in Los Angeles which was the world's major manufacturer of DDT. The Los Angeles sewerage system received the effluent from this factory and, from 1949 until 1971, this resulted in the discharge of 216 t year^{-1} of DDT residues to the sea through the ocean outfall of this sewerage system. A survey made in 1972 suggested that 20 t of DDT residues were trapped in the upper 30 cm of the bottom sediment over an area of 50 km^2 around the outfall (Fig. 6.2).

Discharges from the factory were considerably reduced after 1971 but contamination of bottom-feeding fish in the area continued for several years (Table 6.3). Eventually, the discharge ceased altogether.

Sewage sludge may also contain appreciable quantities of chlorinated hydrocarbons and, if dumped at sea, represents an additional source of contamination. Sewage sludge from Glasgow dumped in the Firth of Clyde in the 1960s made a contribution of 1 t year^{-1} of PCBs to the bottom sediments (Fig. 6.3) until the discharge of PCBs was brought under control.

FATE IN THE SEA

Chlorinated hydrocarbons are extremely insoluble in water, with a saturation concentration of no more than 1 ppb, but they are soluble in fats and adsorb strongly on to particles. Their distribution in the sea is, therefore, far from uniform.

The surface layer of the sea is a film varying from a few μm to 1 mm in thickness. It is extremely difficult to study and is still an area

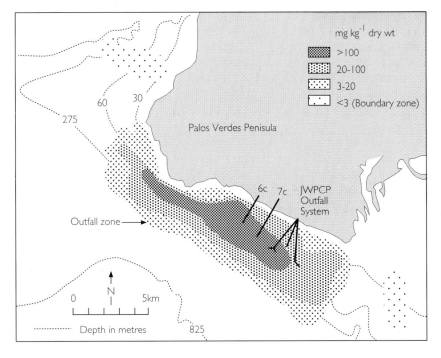

Fig. 6.2 DDT concentration in the top 5 cm of sediment (mg kg^{-1} dry weight) near the outfall of the Los Angeles sewerage system, October 1983. (*Elsevier Science*)

Table 6.3 Percentage of Black Perch (*Embiotoca jacksoni*) which feeds in mid-water, and Dover Sole (*Microstomus pacificus*), a bottom feeder, in samples from the area of the Los Angeles sewage treatment outfall, containing more than the permitted limit of 5 ppm total DDT.

	Black Perch (mid-water)	Dover Sole (bottom)
1970 autumn	75%	—
1971 spring	100%	50%
1971 autumn	71%	80%
1972 spring	73%	86%
1972 autumn	85%	80%
1973 spring	92%	67%
1973 autumn	46%	100%
1974 spring	59%	85%
1974 autumn	63%	82%
1975 spring	55%	83%
1975 autumn	44%	—
1976 spring	23%	—
1976 autumn	18%	—
1977 spring	20%	—

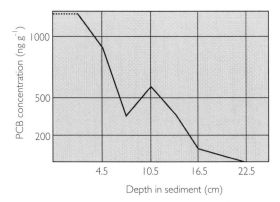

Fig. 6.3 Concentrations of PCBs at various depths in the bottom sediment near the Firth of Clyde sludge dumping ground. (*Elsevier Science*)

of considerable ignorance, but it is known to contain fatty acids. Because of their lipid solubility, organochlorines may therefore accumulate in it. While the total quantity may not be great, the organochlorine enrichment of the surface film may be of considerable importance to surface-living organisms or to birds, such as petrels, that skim fat off the surface of the sea. Since the sea surface is a site of interchange with the atmosphere, organochlorines may be transferred to the air in aerosol droplets.

Where water masses with different physical or chemical characteristics meet, a front is formed. Oceanic and coastal fronts accumulate floating material, including surface oil. Fronts have a high productivity and attract a wide range of fish, birds, and sea mammals, which therefore receive a diet enriched in organochlorines.

A considerable quantity of organochlorines is adsorbed on to particles or on to micro-organisms such as diatoms. This creates problems for analysis because, apart from the difficulty of distinguishing between the various

halogenated hydrocarbons, there is the difficulty of deciding what is incorporated in organisms and what is adsorbed on the outside of them; the former may affect them, while the latter cannot but is biologically available to animals that feed on the contaminated organisms.

Chlorinated hydrocarbons adsorbed on to inorganic particles may ultimately be carried to the seabed which then acts as a sink for these compounds (see Fig. 6.3). However, suspended or resuspended particles, if they are of a suitable size or density, are commonly ingested by filter-feeding animals, and chlorinated hydrocarbons may enter food chains by this route.

Halogenated hydrocarbons, most particularly DDT and its derivatives, now occur in all organisms in all environments, and a considerable quantity in total of the halogenated hydrocarbons in the sea is in the bodies of marine organisms and will continue to circulate within the food webs.

BIOLOGICAL EFFECTS OF HALOGENATED HYDROCARBONS

Problems of analysis

As with petroleum hydrocarbons, halogenated hydrocarbons embrace an enormous variety of related compounds: it may be possible to distinguish 100–150 different organochlorines in a single sample and to identify and measure most of them. Such an abundance of related compounds makes it very difficult to identify which of them is responsible for an observed biological effect. The problem is probably worst for PCBs because the commercial products are mixtures, the exact composition of which varies with the degree of chlorination, the manufacturer, and even with the production batch. Up to 70 PCB compounds may be present in a single sample and the analyst can either estimate total PCBs in the mixture or painstakingly identify and measure individual constituents. The latter approach is very expensive in time and money, so correspondingly fewer samples can be processed. Besides which, unless it is known or suspected which of the many compounds is responsible for the biological effect, nothing is gained by the apparently great accuracy of the detailed analysis. The alternative use of cruder measurements of total PCB contamination may be responsible for discrepancies between reports of the biological effects of PCBs. Some other classes of organochlorines, particularly toxaphenes, present the same problem, but much less attention has been paid to them.

A second problem is more fundamental. By 1965, analytical techniques were able to distinguish DDT from its metabolites DDE and DDD, and from dieldrin, but it was not until 1966 that PCBs were first identified and a year or two later that their separation from DDT was practised in most analytical laboratories. PCBs had been in use and released to the environment for some thirty years, and they were probably present in all samples. Since PCBs can interfere with the detection and measurement of other organochlorines, many, perhaps most, early records of DDT contamination are incorrect. As late as 1980, toxaphenes, which are also widespread contaminants, were not reliably detected in analyses of chlorinated hydrocarbons. Until recently, it was not possible to identify dioxins and furans reliably. Many chlorinated organic compounds have still not been identified and may well be biologically active. Caution must

therefore be exercised in interpreting the effects of halogenated hydrocarbons.

Storage and bioaccumulation

There is evidence that chlorinated hydrocarbons are difficult to excrete and hence they tend to accumulate in the body. Because they are lipid-soluble, they tend to occur in much higher concentrations in fatty tissues than in other tissues. This introduces two risks.

• In times of poor feeding, animals mobilize and use their fat reserves, so increasing the concentration of chlorinated hydrocarbons circulating in the body, possibly to a dangerous level.

• As with other bioaccumulating contaminants, there is a strong possibility of transmission through food-webs and biomagnification, as, indeed, has been established in a number of animals.

Because of the tendency of chlorinated hydrocarbons to be sequestered in fatty tissues, care is needed in comparing levels of contamination in different organisms. Different amounts and concentrations are likely to occur in fat tissue, muscles, gonads, and so on, and even if the total body burden of chlorinated hydrocarbon is known, this has quite different implications for a fat animal than for an emaciated animal.

Accumulation rates for organochlorines, measured by residues in the body divided by residues in the food or the environment, vary widely between species. Figure 6.4 shows the average concentrations of PCBs and six organochlorine pesticides in five species living in the contaminated Weser estuary on the North Sea coast of Germany. The actual concentrations acquired relate to the lipid content of the animal and its position in the food chain, and the sole (*Solea solea*), with at least twice as much lipid as the invertebrates, as well as being a carnivore, is the most contaminated as a result. However, while the sole, shrimp (*Crangon*), and lugworm (*Arenicola*) retain a higher proportion of PCBs than pesticides, the cockle (*Cerastoderma*) acquires relatively little PCB

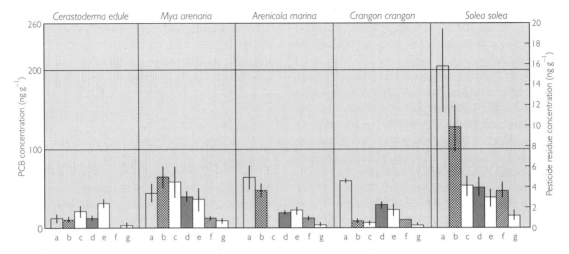

Fig. 6.4 Average concentration of organochlorines in littoral animals in the Weser estuary. (*a*) PCBs, (*b*) DDT, (*c*) dieldrin, (*d*) α-HCH, (*e*) γ-HCH, (*f*) DDE, (*g*) eldosulphan. (*Elsevier Science*)

but proportionately much more γ-HCH—which is not preferentially accumulated by the others. *Crangon* is remarkable in accumulating proportionately less DDT and DDE than the other species. Each species, in fact, has its own profile of organochlorine accumulation.

Even within a species there may be considerable variation in the amount of organochlorine that is acquired. Mussels (*Mytilus* spp.) vary greatly in the DDT residues and PCBs that they contain. A number of different factors undoubtedly contribute to this variation; not all are understood but they include age, tidal level at which they live, sex, stage in the annual breeding cycle, as well as the concentration of organochlorines to which the animal is exposed. It is not only molluscs that show such unexplained intraspecific variations. In the grey seal (*Halichoerus grypus*), males acquire more DDT residues and PCBs than females, but females acquire more dieldrin than males.

The distribution of organochlorines within the animal is not uniform. Figure 6.5 shows how DDT is distributed in different tissues of the sole (*Solea solea*) after receiving ^{14}C-labelled DDT in the diet for four weeks. The brain and, more understandably, the liver both acquire the highest concentrations and eliminate them most slowly.

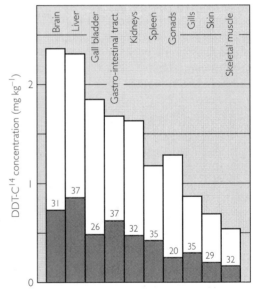

Fig. 6.5 Average distribution of DDT labelled with ^{14}C in tissue of the sole (*Solea solea*) after being dosed for four weeks. Open columns, three days afterwards; filled columns, two months afterwards. Figures are the percentage of ^{14}C DDT retained after four months. (*Springer-Verlag*)

With so much specific and individual variation it is difficult to generalize, but the highest concentration factors (for DDT) are found in bivalve molluscs, such as oysters and clams, where they may reach 70 000. For crustaceans

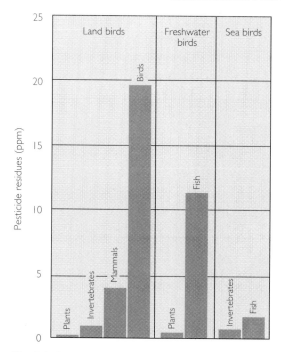

Fig. 6.6 Organochlorine pesticide residues in land, freshwater, and sea-birds, in relation to their predominant diet.

and fish, accumulation factors range between 100 and 10 000; for sea-birds it is 10 or less.

Since organochlorines are eliminated slowly from the body, they may be expected to show biomagnification in food chains. Figure 6.6 shows the average concentrations of organochlorines in land birds, freshwater birds, and sea-birds, in relation to the main types of food they eat. Those with most are land birds feeding on other birds (that is, hawks); they live in an environment where pesticides are most immediately available and they are at the top of a food chain.

Biological effects

It is difficult to be precise about the effect of halogenated hydrocarbons on plants and animals. Because of the low solubility of these substances in water, there is considerable uncertainty about the dose actually received by aquatic organisms in laboratory tests unless the organochlorine is administered by mouth. Because organochlorines are stored in the

fatty tissues of the body, they become biologically available and exert their influence only when the fat tissues are metabolized. Animals may therefore acquire a considerable body burden of halogenated hydrocarbons, but show no ill effects except in conditions of starvation, when fat reserves are mobilized. These facts make laboratory experimentation difficult, but if confirmation of laboratory findings is sought in natural populations there is the additional complication that wild animals are usually contaminated with a variety of halogenated hydrocarbons and it is impossible to tell which one is responsible for whatever symptoms have been detected.

In laboratory cultures of whole phytoplankton from the Caspian and Mediterranean seas, DDT and PCBs reduce primary production (Fig. 6.7) by as much as 50 per cent at a concentration of only 1 μg l^{-1} (ppb)

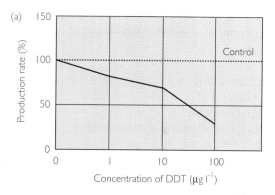

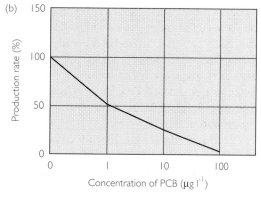

Fig. 6.7 Production (as percentage of control) of (a) whole phytoplankton from the western Caspian Sea exposed to DDT, and (b) whole phytoplankton from the Mediterranean exposed to PCBs.

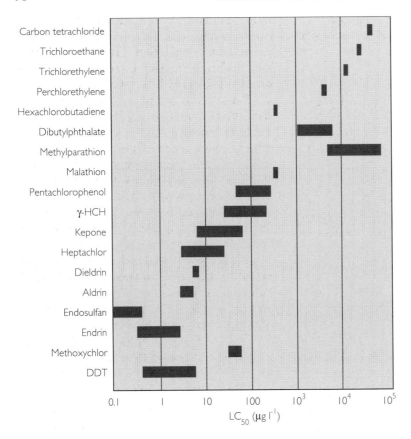

Fig. 6.8 Toxicity of some organochlorines to fish.

in the case of PCBs. Marine fish also appear to be very sensitive to a number of organochlorines (Fig. 6.8). The 96 h LC_{50} for DDT ranges between 0.4 and 89 μg l^{-1}, and for dieldrin between 0.9 and 34 μg l^{-1}, for a variety of teleosts: puffer (*Sphaeroides maculatus*), killifish (*Fundulus heteroclitus* and *F. majalis*), mullet (*Mugil cephalus*), eel (*Anguilla rostrata*), Silverside (*Menidia menidia*), and bluehead (*Thalassoma bifasciatum*). The values for the shrimps *Crangon* and *Palaemonetes*, and the hermit crab *Pagurus longicarpus*, are 0.6–6.0 μg l^{-1} for DDT and 7–50 μg l^{-1} for dieldrin. Bivalve molluscs, on the other hand, with their ability to concentrate organochlorine pesticides without coming to harm, have a 96 h LC_{50} greater than 10 000 μg l^{-1}.

Since halogenated hydrocarbons are known to be subject to bioaccumulation and biomagnification, most attention has been paid to their effect on animals high in the food chain. Very large doses of PCBs and DDT amounting to several g kg^{-1} of the body weight, are required to cause the death of mammals or birds by a single administration by mouth. Such doses are clearly not found in the natural environment, where animals are exposed to much lower levels of contamination of their food, but are continuously, or chronically, exposed to them.

Two sublethal effects of halogenated hydrocarbons in birds and mammals appear to have some ecological significance. DDT and its residues interfere with calcium metabolism and result in egg shells being unusually thin. In both birds and mammals, PCBs have a more direct effect on reproduction, and the eggs of birds contaminated with PCBs, show reduced hatchability. PCBs included in the diet of laboratory rats and mice cause a reduction in the number of offspring produced and poorer survival of those young that are born. Mink are very severely affected in this respect when fed PCBs.

ENVIRONMENTAL IMPACT OF HALOGENATED HYDROCARBONS

Warning signals

During the 1960s there were increasing signs that the widespread and intensive use of pesticides, most obviously DDT, was having unforeseen and unwelcome consequences for the natural environment.

● In 1964, fish were found to be dying in an area around the marine outfall of a Danish factory manufacturing the pesticide parathion. Lobsters, (*Homarus gammarus*) were affected over a much wider area. The factory effluent was found to be lethal to lobsters at a dilution of 1:50 000.

● The Laguna Madre was a productive lagoon system on the coast of Texas, but it became heavily contaminated with pesticides from neighbouring agricultural land. The fish catch in monitoring samples fell abruptly after 1964–5 (Fig. 6.9) and this was attributed to pesticide poisoning.

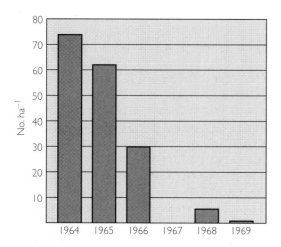

Fig. 6.9 Annual sample catch (in No. ha^{-1}) of sea trout in the Laguna Madre, Texas.

● The Salinas Valley is one of the richest agricultural areas in California, specializing in vegetable and salad crops. The Salinas River which drains the valley does not flow in summer but is blocked by a sand bar at its mouth. During the rainy season, December–April, the river breaks through the bar and flows into Monterey Bay. Although it is wide open to the Pacific Ocean, the bay has a sluggish and irregular circulation and any pollutant remains there for some time. In exceptionally wet seasons the river may burst its banks and flood large areas of farmland. This happened in February 1969, when the river burst the sand bar and flood water carrying enormous quantities of silt flowed into the bay. During the previous 10 years, DDT had been used intensively (it is estimated that 50 t year^{-1} had been used on crops in the valley) and pesticide residues were carried with the silt-laden flood water into the bay. This was followed by exceptionally large numbers of sea-bird deaths in the following months. All had remarkably high liver concentrations of DDT residues:

Brandt's cormorant	107–155 ppm
Western grebe	199–222 ppm
Fork-tailed petrel	313 ppm
Ashley's petrel	412 ppm
Ring-billed gull	805 ppm

In addition, sea lions (*Zalophus californianus*) were found dead or dying on beaches. They had liver concentrations of DDT up to 89 ppm.

The dangers of a widespread ecological impact from the use of persistent pesticides were expressed by Rachel Carson in *Silent Spring*, published in 1963. This book presented the case against pesticides and if some of the predictions were overstated, it drew public attention to the problems associated with the widespread and indiscriminate use of pesticides.

Population decline of predatory birds

The most reliable evidence of the damaging effect of organochlorine pesticides on wildlife was demonstrated in 1967 by Dr D. A. Ratcliffe of the Nature Conservancy in the United

Kingdom. The peregrine falcon (*Falco peregrinus*) was protected in Britain after 1945 and showed a dramatic increase in numbers until, in the mid-1950s, the population went into a sharp decline. This proved to be due to reproductive failure: birds were laying eggs with abnormally thin shells and a large proportion of them were broken during incubation. Examination of museum specimens showed that egg shell thinning had started about 1947. Thinning of egg shells can be produced in ducks and chickens if DDT is included in their diet. High concentrations of DDT residues occurred in peregrines and the yolk of their eggs during the mid-1960s. There was no doubt that DDT was the cause of the population decline of these birds and, with the cessation of the use of DDT for agricultural purposes in Britain, peregrine numbers have increased to their former level.

Pesticides have persisted in the environment, however, and for some reason there has not been a marked decline in DDT residues in sparrow hawks (*Accipiter nisus*) in Britain. This species appears to have been more affected by the cyclodiene organochlorines, and the recovery of its population was delayed until the 'drin' pesticides went out of use in 1975.

The peregrine falcon is a top predator and therefore particularly vulnerable. Comparable effects of organochlorine pesticides were not found in herbivorous or insectivorous birds, but other predators were affected. The North American sparrow hawk (*Falco sparverius*) is related to the peregrine and suffered a similar fate. A number of fish-catching predators were also affected, including the American bald eagle (*Haliaetus leucocephalus*), osprey (*Pandion haliaetus*), and brown pelican (*Pelecanus occidentalis*), all of which suffered a serious population decline. Pelican colonies on the California coast had their first successful breeding in 1972 after nearly 10 years of reproductive failure, following curtailment of the use of DDT as an agricultural pesticide. The situation in these pelicans, however, was complicated by the scarcity of sardines, their staple food, during the breeding seasons of

1969–72 and it is not known how far this contributed to their reproductive failure.

Smaller, fish-eating sea-birds appear not to have been so seriously affected by organochlorine pesticides. One example to the contrary is that of Sandwich terns (*Sterna sandvicensis*) and eider duck (*Somateria mollissima*) on the island of Griend off the coast of the Netherlands. It is thought that effluents from a chemical factory near Rotterdam producing dieldrin, endrin, and telodrin, were responsible for a decline in the size of the tern colony from 20 000 to 650 in 1965. There were also many eider deaths between 1964 and 1967 when measures were taken to prevent the discharge of effluent from the factory. By 1974, the numbers of breeding terns had risen to 5000.

The difficulty in interpreting sea-bird deaths is illustrated by the wreck of sea-birds (mainly guillemots, *Uria aalge*) in the Irish Sea in autumn 1969. Over 12 000 birds came ashore in a dying condition in the space of a few weeks. Some estimates put the death toll at 50 000, or even 100 000 birds. All were emaciated and those autopsied had liver and kidney lesions similar to those caused by PCBs. While the total body burden of PCBs was about the same in the affected birds as in other, apparently healthy, birds, very much higher concentrations were recorded in the liver. This suggests a redistribution of chlorinated hydrocarbon within the body resulting from the loss of fat in these emaciated birds. These results were based on a small sample of birds which proved to be highly variable, with PCB concentrations in the liver ranging from 8–880 μg g^{-1}. Although PCBs may have been a contributory cause of death in some birds, it cannot be concluded that poisoning by chlorinated hydrocarbons caused the wreck. Furthermore, sea-bird wrecks were recorded from time to time in the last century, long before the introduction of these compounds.

Impact on sea mammal populations

Chlorinated hydrocarbons, more particularly PCBs, have been held responsible for the decline of some seal and sealion populations, and may also have affected otters in the Baltic Sea. The evidence for this is far from complete, however, and there are a number of unexplained discrepancies.

Seal populations in the Baltic have been declining since the beginning of this century. Much of the decline was a result of over-hunting, but it has continued despite protection of the seals, and there is a strong suspicion that high levels of PCBs in the animals have been responsible for a failure of reproduction. Since about 1970, only 28 per cent of mature female ringed seals (*Phusa hispida*) have become pregnant each year, instead of the usual pregnancy rate of 80 per cent. Non-pregnant females contain an average of 77 ppm PCBs in the blubber, compared with 56 ppm in pregnant females. About 50 per cent of females have one or both uterine horns blocked by inclusions, effectively rendering them sterile. Some common or harbour seals (*Phoca vitulina*) and grey seals (*Halichoerus grypus*) found dead in the Baltic have shown the same pathological conditions; non-pregnant females of these casualties contained 110 ppm PCBs and pregnant females 73 ppm PCBs in the blubber.

There is no clear evidence that this pathological condition of the uterus is caused by PCBs. Harbour seals on the Dutch coast and grey seals on the north-east coast of England, with higher concentrations of PCBs than those found in the Baltic, do not show this condition. However, there is growing evidence that in mammals, PCBs interfere with ovulation and development, and with the metabolism of thyroid hormones, although this has not been verified experimentally in seals.

PCB contamination may therefore explain why the pregnancy rate of Baltic seals without uterine occlusions is only 50 per cent, and PCBs have been linked with impaired reproduction of harbour seals in the Dutch Wadden Sea. The numbers of otters (*Lutra lutra*) on Baltic coasts have also declined. They feed in inshore waters and the reduction in their numbers has been attributed to the PCB contamination of their preferred food, cyprinid fish.

Outside the Baltic and North Sea, PCB concentrations in the lipids of California sealions (*Zalophus californianus*) have been suggested as the cause abortions, but these specimens had exceptionally high concentrations of DDT residues as well (up to 5077 ppm in the blubber) and cannot be compared with the Baltic seals.

Whatever the suspicions about the harmful effects of PCBs on reproduction, it has to be noted that, at least in the North Sea, seal populations have increased despite sometimes high levels of contamination (Fig. 6.10). Grey seal births have been increasing at an accelerating rate for at least the last 30 years, and harbour seals, after some decline in the 1970s, generally showed a dramatic increase in numbers in the 1980s.

In 1988, an epidemic of viral distemper severely reduced the population of harbour seals in the North Sea. The epidemic started on the Danish and Dutch coasts and then spread to the smaller British colonies. Grey seals were little affected. There has been some speculation that exposure to halogenated hydrocarbons reduced the resistance of these

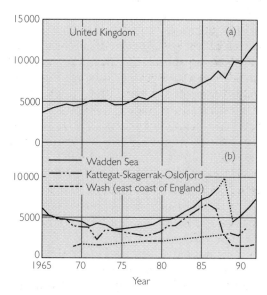

Fig. 6.10 (*a*) Estimated pup production of grey seals in the UK, and (*b*) number of harbour seals in the North Sea.

animals to disease, but there is no firm evidence of this, and lethal epidemics have been observed in seal populations in other parts of the world at various times.

Oestrogenic activity

Some PCBs, pesticides, and a variety of non-halogenated organic compounds are now known to show oestrogenic activity.

Oestrogens are female sex hormones which are involved in regulating the development of sex organs and the control of the reproductive cycle. In mammals, sex is genetically determined, but during development it is reinforced by the hormones produced by the ovaries or testes; exposure of males to oestrogens may therefore interfere with sexual development or performance. In other vertebrates, individuals develop as females if exposed to oestrogen, regardless of their genetic sex, and in that case genetic males may develop ovaries and produce eggs.

The DDT breakdown products op-DDT, pp-DDT, and DDE, as well as the insecticides diethylstilbestrol (DES), chlordecone, methoxychlor, and 1-hydroxychlordene, all display oestrogenic activity. A spillage of dicofoil, an insecticide related to DDT, contaminated Lake Apopka, near Orlando, Florida, in 1980. In subsequent years, a commercially important game fishery disappeared and the alligator population plummeted. Even seven years later, 75 per cent of alligator eggs failed to hatch, compared with 5 per cent in un-

contaminated areas, and those that did hatch produced individuals with sexual abnormalities: 90 per cent of the male progeny did not produce the male hormone testosterone. The alligators contained DDE, a breakdown product of both dicofoil and DDT, at concentrations around 0.1 ppm. This is not sufficient to cause toxic effects, but it is sufficient to disrupt the endocrine system.

A group of substances that have caused particular concern are the alkylphenols. These are phenolic compounds but they do not contain halogens. They are widely used in detergents, plastics, and rubbers and are commonly present in sewage effluents. In one investigation, male trout held in rivers downstream of inputs of sewage effluent developed large quantities of vitellogenin. Vitellogenin is an egg yolk protein which is synthesized in the liver under the control of oestrogen; the enzyme systems that produce it are present in both male and female fish but are normally activated by oestrogen only in maturing females. After only three weeks' exposure, the males had 10000 times the yolk protein normally present in males and as much as is found in a fully mature female (Fig. 6.11).

Other oestrogenic substances have been implicated in the 'feminizing' of Florida panther (Felis concolor coryi), an endangered subspecies of the American puma; males produce less sperm than usual and the sperm tend to be abnormal. Beluga whales (Delphinapterus leucas) in the St Lawrence River in

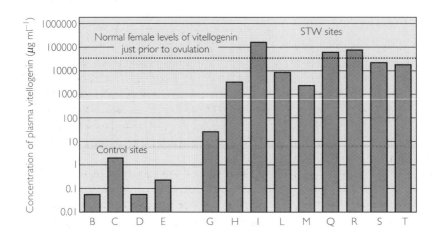

Fig. 6.11 Concentration of vitellogenin in male trout at control sites and after three weeks exposure to effluent from sewage treatment works (STW sites).

Canada, fish and bald eagles (*Haliaetus leuco-cephalus*) in the Great Lakes, and pallial sturgeon in the Missouri and Mississippi rivers have all been shown to be experiencing reproductive difficulties and substances with oestrogenic activity have been implicated in each case. Siberian sturgeon (*Acipenser baeri*) in fish farms are affected by steroids in the commercial feeds.

There is a possibility that the falling fertility of humans in developed countries, and decreasing sperm counts may be the result of exposure to oestrogenic substances that have been identified in materials as diverse as a soybean drink, commercial pet food, pelleted sewage sludge, and substances leaching from plastic tubing.

Restriction of the use of substances with oestrogenic properties has so far been directed at alkylphenol ethoxylates (APEs). These non-ionic surfactants have been widely used as detergents since the 1940s and there is concern about their persistence and bioaccumulation. Because of this they are restricted to industrial use, but it is estimated that 37 per cent of the UK production of 18 000 t year^{-1} eventually reaches the aquatic environment. There has now been agreement through the Paris Commission that their use in cleaning applications will be phased out by the year 2000. APEs will continue to be used in pesticides, paints, and plastics.

THREAT TO HUMAN HEALTH

Dioxins are commonly held to be among the most toxic chemicals known. It is true that the LD$_{50}$ for guinea pigs is 1 μg kg^{-1} body weight, an extraordinarily low figure, but the lethal dose for other rodents is much higher; for hamsters the LD$_{50}$ is 5000 μg kg^{-1} body weight. TCDD results in fetal loss or abortion in rats at doses of 1–10 μg kg^{-1} body weight, and causes developmental abnormalities in mice, such as cleft palate and liver deformities.

Despite many claims to the contrary, the evidence that dioxins are seriously damaging to humans is surprisingly inconclusive. Severe occupational or accidental exposure, such as that following the Seveso incident, when an explosion at a pesticide factory showered the surrounding area with dioxins, resulted in chloracne, minor but reversible nerve damage, and some impaired liver function in the most exposed casualties, but there were no more serious consequences. Studies in Sweden and the United States have suggested a link between dioxins and soft tissue sarcomas, but these cancers have been very rare and other studies have not confirmed the link. Similarly, claims of adverse effects on reproduction among Vietnam veterans exposed to dioxins in herbicides have not been confirmed, and no reproductive ill effects were observed among the Seveso victims in Italy.

In 1968, rice oil contaminated with PCBs caused an outbreak of 'yusho' disease in Japan. In all, 1200 people were affected and suffered darkening of the skin, enlargement of the hair follicles, and eruptions of the skin resembling acne. A majority developed respiratory difficulties which persisted for several years. The rice oil was found to contain 2000–3000 ppm of a 48 per cent chlorinated hydrocarbon and the minimum dose that produced symptoms of the disease was about 0.5 g of PCBs ingested over 120 days, or about 0.07 mg kg^{-1} body weight per day. Similar symptoms to those of yusho disease have been observed in workers at a Japanese condenser factory. This was thought to be owing to local contact with PCBs and the symptoms disappeared when the use of PCBs ceased. It is not entirely clear, though, that PCBs alone were responsible for the yusho incident because workers in Finland occupationally exposed to PCBs and having high concentrations of PCBs in their blood and body fat showed no sign of adverse effects. There have been no confirmed cases of illness resulting from ingesting PCB residues from marine organisms used as food.

Even before the outbreak of yusho disease in Japan, owing to the demonstration of the harmful effects of DDT on top predators among birds there was naturally concern about human exposure to organochlorine pesticide residues in food. Although there was no evidence of harm to humans from this source, legal limits to contamination of foodstuffs in commerce were introduced in a number of countries as a precautionary measure. In the United States this was set at 5 ppm DDT residues and 0.3 ppm dieldrin. There was naturally consternation when it was discovered that human milk of some nursing mothers in Wisconsin (and no doubt elsewhere as well) exceeded that limit for DDT. Had it been cow's milk, it would have been condemned as unfit for human consumption.

Controls on the use of chlorinated hydrocarbon pesticides quickly followed in Europe and North America. Dieldrin and related insecticides were banned or phased out in the early 1970s and DDT had been subject to a progressive withdrawal, particularly for use in agriculture.

This reduction in the use of chlorinated hydrocarbon pesticides was primarily to protect predatory birds and seals which were certainly damaged by them. These substances had not been known to have caused any human deaths, but it was obviously a wise precaution to reduce human exposure to them. However, the replacement of organochlorines by other pesticides did not have a happy record. In 1970, the progressive withdrawal of organochlorine pesticides in the United States reached North Carolina, where the use of DDT on tobacco crops was prohibited for the first time. Parathion, an organophosphorus pesticide which is not persistent but is a potent nerve poison for humans, was widely substituted for DDT. In that season there were 40 human deaths in North Carolina from the careless use of parathion, for which it is necessary to use face masks and protective clothing. Even today, deaths are regularly reported in Nigeria from the careless use of the pesticides that have replaced the organochlorines.

Dieldrin, DDT, and γ-HCH, much of them manufactured in the developed countries, continue to be used on a large scale in the developing world, particularly in India and China. As a result, γ-HCH residues in human milk are higher in India than anywhere else in the world. Whether or not this is harming the population has not been recorded.

RADIOACTIVITY

NATURE OF RADIOACTIVITY

The nucleus of an atom contains **nucleons**, which are either positively charged **protons** or electrically neutral **neutrons**, bound together by powerful nuclear forces which overcome the electrostatic repulsion between the protons. The nucleus is surrounded by a cloud of orbiting **electrons**, each of which carries a negative charge equal to the positive charge on a proton. Normally, the negative charges on the electrons balance the positive charges on the protons, and the number of electrons, and hence the charge on the nucleus, determines the element to which the atom belongs and its chemical properties (Fig. 7.1).

An atom commonly loses or gains one or more electrons in the course of a chemical reaction, or through physical processes. It then becomes a positively or negatively charged **ion** and is chemically much more reactive than the electrostatically neutral atom. The process by which this occurs is **ionization**.

Atoms of the same chemical element have the same number of protons in the nucleus, but the number of neutrons may vary. These variants, known as **isotopes**, have the same chemical properties but differ in their nuclear mass. Thus, 99 per cent of naturally occurring carbon has a nucleus containing 6 protons and 6 neutrons and is designated carbon-12 or ^{12}C because its nucleus contains 12 nucleons; but 1 per cent contains 7 neutrons and is designated

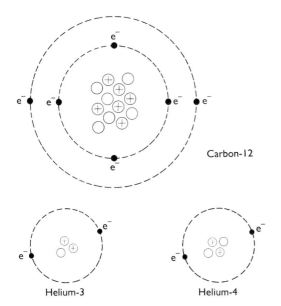

Carbon-12

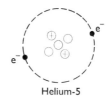

Helium-3   Helium-4   Helium-5

Fig. 7.1 The structure of some atomic nuclei.

carbon-13. Both of these are stable isotopes, but some other isotopes are unstable. Potassium, with 19 protons in the nucleus, normally occurs in a form with 20 neutrons as potassium-39; potassium-40, with 21 neutrons, also occurs naturally, but is unstable. The instability of the nucleus is remedied by a change in the ratio of protons to neutrons, accompanied by the emission of particles and energy. This is known as **radioactivity**, and the unstable forms are **radioisotopes** or **radionuclides**. Each radioisotope has its characteristic nuclear change and form of emissions.

α radioactivity

The unstable nucleus emits an α particle consisting of two protons and two neutrons. Since the nucleus loses two positive charges, the atom becomes that of an element two places lower in the periodic table. α particles are relatively slow moving and lose their energy in a short distance; they are stopped by a few centimetres of air or only 40 μm of tissue. They are, however, intensely ionizing in the matter through which they pass and can cause more damage to living tissue than particles with a longer path. Nuclei emitting α particles are therefore of biological consequence if they are taken into the body, for example by ingestion or inhalation.

β radioactivity

A neutron in an unstable nucleus spontaneously changes into a proton, or vice versa, and the resulting atom is that of an element one place higher or lower, respectively, in the periodic table. To conserve electrostatic charges, a β particle is emitted. This is a negatively charged **electron** if a neutron has changed into a proton, or its positively charged equivalent, a **positron**, if a proton has changed into a neutron. β particles vary widely in their energy but lose most of it within a relatively short distance and can be screened by a few millimetres of perspex or 40 mm of tissue. Like α particles, their biological significance is greatest if a β emitter is taken into the body.

Spontaneous fission

Nuclei of some very heavy unstable elements always have a large excess of neutrons. The nucleus breaks into two large fragments representing elements in the middle of the periodic table, accompanied by a few free neutrons. This is known as **spontaneous fission** and the fission products are themselves unstable. Neutrons are slowed down only by collisions with other nuclei and, because of the infrequency of this, neutrons penetrate matter for a considerable distance. Although they carry no electric charge and do not cause ionization, the nuclei with which they collide cause intense ionization over a short distance (like α particles) and the nucleus that finally absorbs the neutron shows strong γ radiation.

γ radiation

γ rays are similar to X-rays and, like them, are deeply penetrating and strongly ionizing. Living tissues need to be shielded from γ radiation by a considerable thickness of heavy material such as lead or concrete to absorb the radiation. In addition to being emitted by nuclei bombarded with neutrons, some of the energy released by α and β emitters, particularly the latter, is in the form of γ rays.

UNITS

Becquerel (Bq)

Radioactivity is measured by the frequency with which radioactive disintegrations take place in a substance. The becquerel is one nuclear disintegration per second. High levels of radioactivity may reach terabecquerels ($1\ TBq = 10^{12}\ Bq$). The becquerel replaces the old unit of radioactivity, the curie (Ci), which is defined as the amount of radioactivity displayed by 1 g of radium (^{226}Ra) and equals $3.7 \times 10^{10}\ Bq$.

Gray (Gy)

The becquerel takes no account of the nature of a disintegration, merely its frequency. For biological purposes it is more important to know the radioactivity absorbed by a tissue or an organism. This is measured by the gray (Gy), defined as the amount of radiation causing 1 kg of tissue to absorb 1 J of energy. The old unit replaced by the gray is the rad, which is 0.01 Gy.

Sievert (Sv)

Different kinds of radiation cause different amounts of damage to living tissue for the same energy, and neutrons or α particles have about ten times the effect of β or γ particles for the same number of grays. The sievert is an arbitrary unit designed to take account of this difference. Thus, a dose of 1 sievert could be made up of 1 gray of γ particles or 0.1 gray of neutrons. The old unit replaced by the sievert is the rem, which is 0.01 Sv.

Half-life

The radioactivity of a substance declines with the passage of time. After one half-life, the radioactivity is halved. Each radionuclide has its characteristic half-life, which may be a fraction of a second, days, months, or millions of years. The half-life of ^{226}Ra is 1602 years. Radioactivity is inversely related to the half-life and a substance with a long half-life has low radioactivity.

INPUTS OF RADIOACTIVITY TO THE SEA

Background radioactivity

Radioactivity is a natural phenomenon. Seawater is naturally radioactive, largely due to the presence of potassium-40, but it also contains decay products of uranium and thorium, and receives a continuous input of tritium (^{3}H, the radioactive isotope of hydrogen) through the activity of cosmic rays (Table 7.1).

Heavy radionuclides have a low solubility in water and tend to be adsorbed on to particulate matter. They therefore accumulate in sediments; fine sediments, with their large surface area, tend to adsorb more than coarse sediments. Thus, while oceanic seawater has a radioactivity of about 12.6 Bq l^{-1}, marine

Table 7.1 Natural levels of radioactivity in surface seawater

Radionuclide	Concentration (Bq l^{-1})
Potassium-40	11.84
Tritium (^{3}H)	0.022–0.11
Rubidium-87	1.07
Uranium-234	0.05
Uranium-238	0.04
Carbon-14	0.007
Radium-228	$(0.0037–0.37) \times 10^{-2}$
Lead-210	$(0.037–0.25) \times 10^{-2}$
Uranium-235	0.18×10^{-2}
Radium-226	$(0.15–0.17) \times 10^{-2}$
Polonium-210	$(0.022–0.15) \times 10^{-2}$
Radon-222	0.07×10^{-2}
Thorium-228	$(0.007–0.11) \times 10^{-3}$
Thorium-230	$(0.022–0.05) \times 10^{-4}$
Thorium-232	$(0.004–0.29) \times 10^{-4}$

sands have a radioactivity of 200–400 Bq kg^{-1}, and muds 700–1000 Bq kg^{-1}. In parts of the world with very high natural levels of radioactivity, marine sands have correspondingly high radioactivity levels. The best known examples are in Kerala in south-west India and in the provinces of Espirito Santo and Rio de Janiero in Brazil. At one popular bathing beach at Guarapari, near Rio de Janiero, the visitor is exposed to a dose rate of 20 μGy h^{-1}. For comparison, the dose rate on the most heavily contaminated sediments around the reprocessing plant at Sellafield is 0.46 μGy h^{-1}.

Weapons testing

Inputs of radioactivity to the sea from human activities began in the late stages of the Second World War with the explosion of the first nuclear weapons, and continued with nuclear weapons testing until the signing of the test ban treaty between the USA, the former USSR, and the UK in 1963. From then until 1974 atmospheric tests were conducted by France in the Tuamotu Archipelago (at Mururoa and Fantataufa) in the Pacific Ocean, and by the People's Republic of China; these made relatively minor contributions. Since that time, all tests are believed to have been carried out underground with no discharge to the atmosphere.

These nuclear weapons contained enriched uranium and plutonium, and when exploded under water or close to the ground produced more than 200 different fission products and isotopes. These were carried in fine dust into the atmosphere where they circled the globe many times before settling back to earth again as fallout. Many of the radioisotopes produced do not occur naturally; some have a very short half-life, but the longer lived radioisotopes allow the distribution and extent of the contamination from this source to be detected. The most important of these are strontium-90 and caesium-137 with a half-life of about 30 years, and plutonium-239 with a half-life of 24 400 years.

Fallout peaked in the 1960s (Fig. 7.2) and

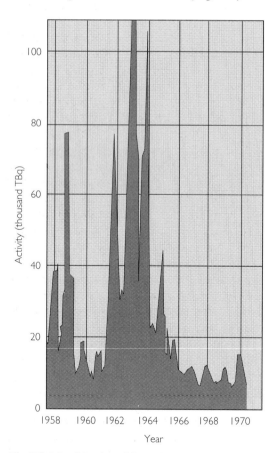

Fig. 7.2 Monthly deposition of strontium-90 in the northern hemisphere derived from weapons testing.

this was reflected in levels of contamination of haddock and cod in the Barents Sea (Fig. 7.3). Because of the global pattern of atmospheric circulation, most of the fallout occurred between latitudes 45°N and 45°S, with higher levels in the northern hemisphere where most of the explosions took place. There has been negligible input from weapons testing since the mid-1960s (Fig. 7.2) and radioactivity from this source has been subject to natural decay ever since then. The average radioactivity of ocean water in the north Atlantic in the early 1970s attributable to fallout is shown in Table 7.2, but its distribution is far from uniform. Initially, the highest levels occurred in surface waters, but radionuclides are transferred from surface to deeper water mainly through biological processes, being carried down in dense faecal pellets of planktonic organisms and the cast exoskeletons of crustaceans.

Comparison of radionuclides in mussels and oysters on the Atlantic, Gulf, and Pacific coasts of America in 1976–8 and in 1990 show that plutonium-239 and -240 and caesium-137 in bivalves decreased significantly (eightfold), reflecting the continuing decline of fallout material. Americium-241 contamination was higher on the west coast of America than on the east coast in both 1976–9 and 1990; this is thought to be caused by upwelling of intermediate Pacific waters associated with the California current.

Liquid wastes

Radioactive substances are included in cooling water and other liquid wastes from nuclear reactors and may be discharged into the sea.

Nuclear-powered ships and submarines release some radioactivity, but the amounts are trivial compared with the discharges from nuclear power stations and fuel reprocessing plants. Furthermore, such discharges as these are widely distributed in the oceans, not confined to one area as in the case with discharges from land-based installations.

By 1991, 497 nuclear reactors for generating electric power were in operation or under construction in 30 countries. Several types of reactor are in use, but all depend on the

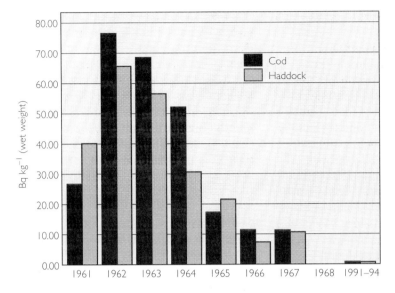

Fig. 7.3 Mean β-activity in the edible part of cod and haddock caught in the Barents Sea, 1961–94.

Table 7.2 Average (and range of) radio-activity in north Atlantic waters in the early 1970s due to fall-out from nuclear weapon tests

Radionuclide	Radioactivity (Bq l^{-1})	
Tritium (^{3}H)	1.78	(1.15–2.74)
Caesium-137	0.008	(0.001–0.03)
Strontium-90	0.005	(0.0007–0.02)
Carbon-14	0.0007	(0.0004–0.015)
Plutonium-239		(0.0001–0.0004)

energy derived from the fission of uranium, usually in the form of uranium dioxide. The core of the reactor containing the fuel rods generates great heat, and a coolant passes through the core to maintain a uniform internal temperature. The coolant may be light or heavy water, carbon dioxide, or molten sodium. The heat is transferred indirectly to water and used to generate steam to drive turbines, which generate electricity in the conventional way. This water is then condensed and recycled, but large volumes of cooling water are required for this process and are then discharged into the sea in the case of coastal or estuarine nuclear power stations.

The discharged cooling water inevitably acquires some radioactivity and the discharge also contains liquid wastes with low radioac-tivity from processes involved in the handling of spent fuel rods from the reactor. In Britain, the total radioactive discharge and the discharge of particular radionuclides are limited by regulations for each nuclear power station, depending on local circumstances. Thus, the nuclear power station at Bradwell on the Blackwater estuary in Essex was originally permitted to discharge (in addition to 55.5 TBq year^{-1} of tritium) 7.4 TBq year^{-1} of radioactivity, but because of the presence of commercial oyster beds in the area, the discharge of zinc-65 (which oysters accumulate) was limited to 0.185 TBq year^{-1}. At other nuclear power stations where there is no such commercial interest, zinc discharges are not specifically regulated. In practice, emissions from nuclear power stations have been dramatically reduced in recent years. Zinc-65 emissions from Bradwell, for example, were 0.0012 TBq in 1992, less than 1 per cent of the previous limit.

The fuel elements remain in the reactor for two or three years, by which time they have lost their efficiency but still contain large amounts of uranium-235 which has not yet undergone fission, together with a variety of other radionuclides, including plutonium-239 and americium-241. Spent fuel rods may be removed for permanent safe storage on land, as is the current practice in Canada and the

United States, or, more commonly, are taken to a reprocessing plant where the uranium is recovered for reuse and the plutonium and fission products are extracted. Major reprocessing plants in Europe are at Sellafield on the north-west coast of England and Cap la Hague, near Cherbourg, on the French Channel coast. There are other reprocessing plants at Dounreay on the north coast of Scotland, Mercoule in France, which discharges into the River Rhône, and at Karlsruhe in Germany. Reprocessing plants also exist in India, Japan, Russia, and the USA. Reprocessing plants discharge large quantities of waste water with a low radioactive content but, because of the quantities involved, they dwarf the output from nuclear power stations. Although the discharges from all the European reprocessing plants have been carefully monitored, more is probably known about the impact of the discharges from Sellafield than about any other source of man-made radioactivity.

Discharges from Sellafield rose to a peak in the early 1970s when the maximum permitted release of α emitters was 222 TBq year^{-1} and of β emitters was 11 000 TBq year^{-1}. Since that time, discharges have been greatly reduced, and by 1994 the maximum permitted discharge of α emitters was 4.7 TBq year^{-1}, and of β emitters was 400 TBq year^{-1}. In practice, actual discharges have been substantially below these limits. The most significant discharges, because of their long half-life, are caesium-137 and -134, ruthenium-106, strontium-90, and americium-241, and separate maximum permitted discharges are set for these and a number of other radionuclides. In recent years there has been some concern about the possibility of the exposure of humans to α emitters, primarily plutonium and americium, and discharge of these has now been reduced to a very low level (Fig. 7.4).

The behaviour of these radionuclides in the sea depends on their chemical form and physico-chemical characteristics. Caesium-137 remains in solution and has a half-life of 30 years. Since the isotope does not occur naturally, it can serve as a marker for Sellafield liquid emissions, showing their course and dilution in the Irish Sea (Fig. 7.5) and round the north of Scotland into the North Sea.

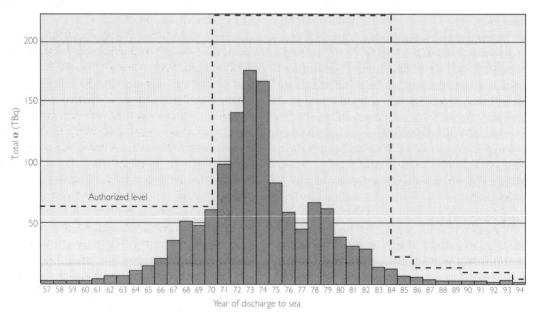

Fig. 7.4 Permitted and actual discharge of α radioactivity to the sea from the Sellafield reprocessing plant. (*Publishing with the permission of the Controller of Her Majesty's Stationery Office.*)

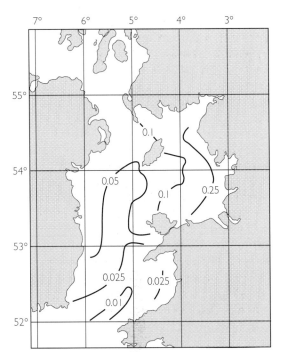

Fig. 7.5 Concentration (in Bq kg^{-1}) of caesium-137 in filtered surface water from the Irish Sea, November 1993.

Other radionuclides, particularly those of ruthenium and plutonium, tend to adsorb strongly on to particulate matter and so are carried to the seabed where they accumulate in fine substrata. Some seabed sediments in the area influenced by Sellafield are contaminated with up to 9.1 Bq g^{-1} of plutonium-241. Resuspension of particles and the local hydrography results in some accumulation of radioactive particles in estuarine muds and salt marshes on the Cumbrian and Solway coasts and this has to be taken into account in calculating the radiation exposure of fishermen in the area who regularly work on the mud-flats.

Solid wastes

Sea dumping of radioactive solid wastes began in 1946, but was progressively brought under international control and is now no longer practised. In 1972, under the London Dumping Convention which regulates all dumping at sea, the disposal of high level radioactive wastes at sea was banned, and in 1994 the ban

was extended to include low and intermediate level solid wastes as well.

High level wastes are essentially those that are sufficiently radioactive to generate heat and are defined as those containing, per tonne of material, 37 000 TBq tritium, or 37 TBq β and γ emitters, or 3.7 TBq strontium-90 and caesium-137, or 0.037 TBq γ emitters with half-lives over 50 years. Low and intermediate level radioactive wastes include such material as protective clothing, laboratory glassware, contaminated piping, concrete and building material, and so on, and is derived from nuclear power stations, reactors operated by industry and nuclear research centres, universities, research institutes, and radiological services in hospitals.

For disposal, this waste is packed in concrete-lined steel drums, either embedded in bitumen or resin, or with a pressure equalization device. This ensures that the drum reaches the seabed intact without imploding under the great pressure in deep water. Ultimately the containers will corrode and their contents leach out into the surrounding water, but the delay results in a loss of radioactivity as the short-lived radionuclides decay, and the slow release of the contents ensures great dilution.

Between 1946 and 1970, when sea dumping ceased, the USA disposed of 107 000 canisters containing a total of 4300 TBq at deep water sites offshore. Almost half the disposals were in an area 80 km west of San Francisco in the Pacific Ocean, and other substantial disposals were made at two deep water sites 225 and 354 km off the New Jersey coast. Smaller quantities were dumped at various times in Massachusetts Bay in the Atlantic and at a number of sites in the Pacific Ocean.

The UK started to dump low level solid radioactive waste at a number of sites in the north-eastern Atlantic in 1949; after 1971 this activity was centred at a single site 900 km from the nearest land and 550 km beyond the edge of the continental shelf in 4400 m of water. After 1967, these disposals included wastes from Belgium, France, Germany, Italy, the Netherlands, Sweden, and Switzerland, as

Table 7.3 Amount of intermediate and low level radioactive waste dumped in the north-east Atlantic

Year	α emitters (TBq)	β/γ emitters (TBq)	Tritium (TBq)
1974	15.5	40 700	
1975	28.9	1130	1100
1976	32.6	1200	775
1980	70.3	3075	3630
1981	77.7	2930	2750
1982	51.8	1830	2860
Total 1948–82	660	38 000	15 000

well as from the United Kingdom. The amounts of radioactivity dumped until the use of this site was suspended in 1983 are shown in Table 7.3. Japan planned to dump low level radioactive waste in deep water in the western Pacific, but this scheme was abandoned.

Between 1964 and 1990, the former Soviet Union made intensive use of dump sites in the Barents and Kara seas around Novaya Zemlya (Fig. 7.6), as well as in the Sea of Japan. In addition to 576 TBq of low and inter-mediate level waste, at least one redundant nuclear submarine, including the fuel rods, and seventeen reactors from nuclear sub-marines, seven including spent fuel and ten without, have been dumped in the Kara Sea and fjords on the east coast of Novaya Zemlya.

As on land, ocean disposal sites for radio-active waste need to be selected so that no risk is presented to the human population. There is no danger of drums of waste being accidentally trawled by fishermen at deep water sites. The waste containers will, of course, eventually corrode and release their contents, although one container recovered from the US 2800 m deep dump site in the Atlantic after 15 years showed no sign of de-terioration and was judged to require at least 300 years before it lost its integrity. In that time, much of the radioactivity of the contents will have decayed, but isotopes with a long half-life will be released. Those such as pluto-nium and ruthenium that are virtually insolu-ble in water will be adsorbed on to particles

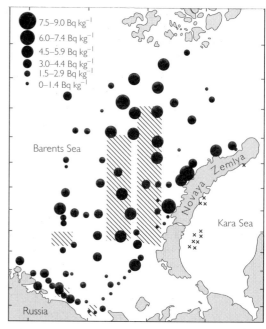

Fig. 7.6 Radioactivity of sediments (in Bq m^{-2}) at waste disposal sites in the Barents Sea; crosses, solid waste disposal sites in the Kara Sea; cross-hatched areas, disposal sites for liquid wastes.

and remain in the sediment at the dump site.

Soluble isotopes will be subject to slow dis-persion by bottom currents and enormous dilution. At the European Atlantic dump site bottom currents on the abyssal plane flow towards the Antarctic, where they surface. Antarctic plankton might therefore be a route by which radionuclides could reach the human population through the food chain. Even if dumping were resumed at ten times the pre-vious rate, the peak annual dose would be less than 10^{-4} mSv, arising 100–500 years after the start of dumping. In hypothetical pathways including the consumption of abyssal fish, the highest individual dose rate is estimated to be 2×10^{-4} mSv year^{-1}. This is an exceedingly low individual dose rate when compared with the ICRP (International Commission on Radiological Protection) recommended limit of 1 mSv year^{-1}.

Surveys made in the Barents Sea by the Norwegian fishery authorities during 1990–3 found that the highest levels of radioactivity in sediments, 150 Bq m^{-2}, occurred in areas where dumping had taken place (Fig. 7.6), and that fish contained 1.6–3.5 Bq kg^{-1} wet weight of caesium-134 and -137. Seaweeds contained 4–10 Bq kg^{-1} dry weight. In monitoring landings of cod from the Barents Sea during the period of atmospheric bomb tests in the 1950s and 1960s, β activity never exceeded 80 Bq kg^{-1} wet weight, even at the most intense period (Fig. 7.3). In the 1990s, no commercial landings of fish exceeded the detection limit of 20 Bq kg^{-1}.

Accidental inputs

A number of accidents have resulted in radioactive materials being added to the sea. These include plutonium in a nuclear powered satellite that re-entered the atmosphere prematurely in 1964, the crash of two US B-52 bombers carrying nuclear weapons (one off the Spanish coast in 1966, the other near Thule in Greenland in 1971), the loss of two US nuclear-powered submarines in the Atlantic in 1963 and 1968, and of six Russian nuclear-powered submarines in the Norwegian and Kara Seas; the most recent being the *Komsomolets* which sank after a fire and explosion in 1989. No harmful effects of these accidents on the marine environment have been detected and, although some attempt is being made to seal the most recent casualty which lies in 1700 m of water in the Norwegian Sea, there is general agreement among oceanographers that the safest policy would be to leave the wreck undisturbed.

These accidents are insignificant beside the consequences of the disastrous fire and explosion at the Chernobyl nuclear reactor in April 1986, which resulted in radioactive fallout over considerable areas of Europe, particularly where there were heavy rainstorms as the nuclear cloud passed over. The sea area most affected was the Baltic Sea, although there were significant inputs in the northern Adriatic, the North Sea, the northwest coast of Scotland, and the Irish Sea. The dominant radionuclide was caesium-137, as with routine discharges from nuclear installations. Land and freshwater environments have more limited dispersal routes than the

sea and a number of land areas remained significantly contaminated ten years after the accident; the fallout had a very small effect in the sea.

Seawater samples were collected all over the Baltic Sea by the German Hydrographic Institute during October and November 1986 and the mean concentration of caesium-137 was then 0.18 Bq l^{-1}, and of caesium-134 was 0.095 Bq l^{-1}. The average caesium-137 content of Baltic water before the accident, in the period 1980–5, was 0.018 Bq l^{-1}.

During the summer following the accident, fish from the western part of the Bothnian Sea contained an average of 64 Bq kg^{-1} of caesium-137 and 31 Bq kg^{-1} of caesium-134. In Finnish waters, fish samples collected between May and November 1986 averaged 41 Bq kg^{-1} of caesium-137 and 19 Bq kg^{-1} of caesium-134. Much lower concentrations were recorded in fish samples from the southern Baltic Sea.

Assuming an annual consumption of 100 kg of Baltic fish per person per year, the most critically exposed group (see Table 7.6, p. 109) would have received a dose of less than 0.08 μSv year^{-1}. This is only one-thirtieth the exposure of the most critically exposed consumers of freshwater fish affected by the accident.

ENVIRONMENTAL IMPACT OF RADIOACTIVITY

It is difficult to identify the ecological impact of radioactivity in the sea. Toxicity tests on marine organisms are not very informative. Most of the field investigations have been conducted in relation to the discharge from the Sellafield reprocessing plant and, since the Sellafield discharges have historically been among the largest in the world, the experience there gives a good indication of the ecological impact of radioactivity in the sea.

Lethal doses

The measurement of acute lethal doses of radioactivity is complicated by the fact that

Table 7.4 Acute lethal radiation doses for adults of marine organisms

Organism	Dose (Gy)	Type of experiment
Blue-green algae	4000–12 000+	LD_{90}
Other algae	30–1200	LD_{90}
Protozoa	<6000	LD_{90}
Mollusca	200–1090	LD_{50}
Crustacea	15–566	LD_{50}
Fish	11–56	LD_{50}

LD_{50} values are based on total mortality observed during 30 days following exposure to radiation.

Table 7.5 Radioactivity of *Porphyra* collected near Sellafield in 1974

Radionuclide	Concentration $(Bq\ g^{-1})$
Ruthenium-106	8.29
Cerium-144	0.90
Zirconium-95/Niobium-95	1.37
Caesium-137	0.093
Plutonium-239, -240	0.051
Americium-241	0.023
Strontium-90	0.016

damage inflicted on test organisms may not be revealed for some time after exposure to radiation, and sublethal genetic damage cannot be detected until the next or later generations. This is less of a problem for microorganisms with a very short lifespan but, for these, measurement of the LD_{50} is hardly appropriate and, instead, the LD_{90} is used. In practice this is a measure of the dose at which almost the entire culture dies. For macroscopic animals, the LD_{50} is based on the mortality occurring within a given time after exposure to radiation. This time is arbitrarily selected, but 30 days is a convenient period and most animals surviving so long stand a good chance of living an appreciable time longer.

The LD values given in Table 7.4 are therefore intended only to give an impression of the susceptibility of different organisms to radiation damage. There is wide variation but, as a rule, the more advanced and complicated the animal, the lower is the radiation dose necessary to cause fatal damage. Gametes and larval stages are more susceptible to radiation damage than adults.

Bioaccumulation and food webs

Radionuclides behave chemically in the same way as their non-radioactive, naturally occurring isotopes, but the possibility of bioaccumulation and biomagnification in food chains has greater significance if the substance accumulated is radioactive.

Algae are able to acquire large concentra-

tions of substances from the surrounding water. *Porphyra umbilicalis*, since it is the only seaweed eaten in quantity by sections of the human population in Britain, has attracted particular attention. In the neighbourhood of Sellafield, where it is exposed to a considerable variety of radionuclides from the marine outfall from the reprocessing plant, it accumulates 10 times the concentration of caesium-137 found in the water, 400 times the concentration of zirconium-95 and niobium-95, 1000 times the concentration of cerium-144, and 1500 times the concentration of ruthenium-106. Concentrations of radionuclides in *Porphyra* collected near Sellafield at the period of maximum discharges in 1974 are shown in Table 7.5. Other seaweeds in the area also accumulate radionuclides, but different elements predominate: *Enteromorpha* and *Ulva* accumulate much higher concentrations than *Porphyra* of zirconium, niobium, and cerium, and also accumulate plutonium-239. Fucoids, although they are rather poor accumulators of ruthenium-106, accumulate very much higher concentrations than *Porphyra* of plutonium-239, and for this reason have proved to be useful monitors of this radionuclide. Despite the high concentrations of radioactive materials in these algae, they are unaffected by them.

Planktonic crustaceans appear to be poor assimilators of radionuclides received in their food and they lose a high proportion of their intake in faeces. The fact that they moult at frequent intervals means that radionuclides adsorbed on to them are regularly shed. Planktonic crustaceans are therefore not an important pathway for radionuclides in pelagic food webs. Most inshore fish are, however, benthic feeders and in an area with a localized input of radioactive material, as at Sellafield, sedimentation processes ensure that there is an accumulation of radionuclides on the seabed with corresponding contamination of the infauna. Crabs (*Carcinus maenas*) accumulate much greater quantities of plutonium-237 than plaice (*Pleuronectes platessa*) when fed on *Nereis diversicolor* labelled with this radioisotope (Fig. 7.7) most of the radionuclide was accumulated in the digestive gland of the crab; in the plaice it was found adhering to the gut wall but not accumulated in other organs.

Bivalves are notorious for their ability to accumulate very high concentrations of metals, although specific abilities vary widely. Scallop (*Pecten maximus*) accumulate large quantities of manganese, oysters (*Ostrea*) large amounts of zinc, and mussels (*Mytilus*), iron. The very small amounts of the radio-

isotopes of these elements included in the discharges are readily taken up by these species.

Exposure levels

Bottom-living fish are likely to be exposed to higher levels of radioactivity than pelagic species because of the adsorption of radionuclides to particles that accumulate in the seabed. Experiments have been conducted to measure the exposure of plaice in the northeast Irish Sea by attaching dosimeters to their ventral surface. Individuals caught up to two and a half years after release proved to have acquired a mean dose of 3.5 μSv h^{-1}, with occasional individuals that had received a dose of 25 μSv h^{-1} (Fig. 7.8). This maximum figure is about the same as the calculated dose based on the seabed radioactivity near the Sellafield outfall. Evidently, the migrations of the fish result in the great majority of them receiving a much smaller dose. The minimum dose at which minor radiation-induced disturbances to physiology and metabolism can be demonstrated in the laboratory is 100 μSv h^{-1}, and it is evident that

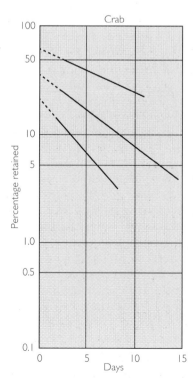

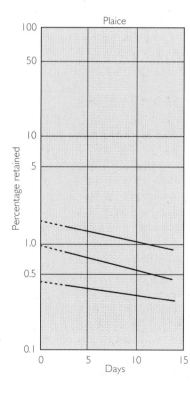

Fig. 7.7 Percentage retention of plutonium-237 by three crabs (*Carcinus maenas*) and three plaice (*Pleuronectes platessa*) after being fed *Nereis diversicolor* containing the radionuclide. (*Macmillan*)

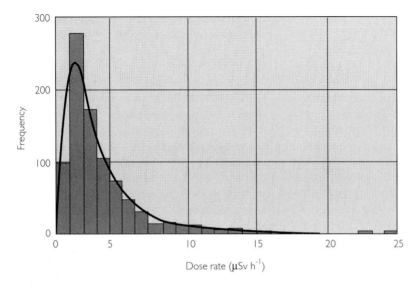

Fig. 7.8 Dose of radiation received by tagged plaice (*Pleuronectes platessa*) in the north-east Irish Sea.

bottom-living fish, even in the most heavily contaminated areas, receive far less than such minimally damaging exposures.

Population effects

Even though marine organisms generally have a relatively high tolerance of radioactivity, it must be expected that damage to some susceptible individuals and genetic disturbances are an inevitable consequence of any level of radiation exposure, whether its source is natural or man-made. This would be manifested in increased mortality of adults or eggs and larvae, but in view of the enormous natural mortality this would be very hard to detect.

In one Canadian experiment, eggs and larvae of the chinook salmon (*Oncorhynchus tschawytscha*) raised in a salmon hatchery were subjected to 5–200 mGy per day, and returns to the river were measured in subsequent years. Large batches of eggs and young were used in the experiments and there was an increased number of abnormalities in young fish at all dose rates, but no decline in the number of adults returning to the river for spawning was ever detected.

Existing levels of radiation in the sea have so far produced no measurable environmental impact on marine organisms or ecosystems. This is not to say that there are no particularly sensitive organisms or ecosystems that might

be adversely affected, but it will evidently require detailed investigations to expose them.

HAZARD TO HUMAN HEALTH

The natural environment appears to be unaffected by present levels of radioactivity, but there is naturally more concern about risks to the human population.

Exposure to radiation

The human population is exposed to radiation from a variety of sources, natural and man-made (Fig. 7.9). Natural sources are responsible for an annual dose of about 2200 μSv year^{-1}, caused by exposure to radon gas seeping from the ground, terrestrial radiation (including building material such as stone or brick), cosmic rays, and sources within the body, mainly potassium-40 ingested in food.

Artificial sources of radiation are almost wholly in medical diagnosis and therapy (chiefly X-rays). the average annual dose from this source is about 300 μSv. Radioactive discharges, fallout, other industrial sources, and miscellaneous sources make a very small contribution to the average annual exposure.

The average exposure figures conceal wide

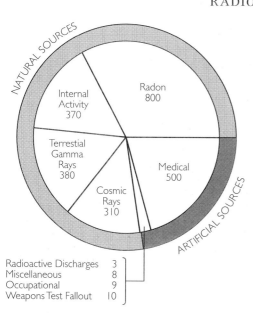

Fig. 7.9 Average annual dose (in μSv) of radiation from natural and artificial sources to which the human population is exposed in the UK.

Table 7.6 Estimated maximum doses of radiation received by critically exposed groups of people

Direct inputs into the marine environment	Annual individual effective dose equivalent (mSv)
Sellafield	0.3–3.5
Dounreay	0.03
Cap de la Hague	0.1
Other nuclear sites	0.0001–0.3*
Solid waste disposal	0.00002†
Chernobyl fall-out	0.08‡
Weapons testing fall-out	0.001–0.01
Naturally occurring radiation	<2

*Doses towards the higher end of the range are largely due to the influence of Sellafield.

†Peak hypothetical dose in the future due to all past disposals.

‡During 1986.

variations. Radon is particularly prevalent in granitic areas where it may accumulate in houses to a dangerous level. People living at high altitudes are more exposed to cosmic rays than those living at sea level, and exposure to 'miscellaneous' artificial sources is almost entirely a result of flying. Medical exposure may vary from zero to near lethal doses.

Exposure to radioactivity from the sea is either through the consumption of seafood that has accumulated radioactive substances, or by external exposure to radioactive sediments on beaches. Some groups of people, because of their diet, work, or location, may receive higher doses than members of the general public. Table 7.6 shows the estimated maximum doses that may have been received by a selection of such critically exposed groups in northern Europe in recent years. Different groups of people are involved in each case. Massive doses of radiation result in radiation sickness and death but, at the low doses described so far, exposure must continue over a long period (chronic exposure) before adverse effects are clearly manifested.

Effects of radiation

Chronic exposure to ionizing radiation may produce **somatic** effects, most importantly leukaemia and cancer of the bone, thyroid, lungs, or breast; or **genetic** damage manifested in some abnormalities or defects in the next generation, resulting from irradiation of the gonads. While these effects undoubtedly occur, it is always difficult to establish a connection between low level chronic exposure and the incidence of an effect in any individual. Such diseases do not usually reveal themselves until long after the period of exposure; there is not a strict relationship between the intensity and duration of exposure and the incidence of disease or effects in the individual at risk; and all these diseases occur in any normal population, whether or not it is exposed to a particular source of radiation. Equally, genetic effects show a dose relationship only at the level of the population and the consequences of exposure for the individual are unpredictable. The nature of the genetic damage resulting from radiation does not differ from genetic defects that occur spontaneously.

Different radionuclides present different kinds of health hazard, depending on their chemistry. In the same way as the stable isotopes, depending on the form in which radioisotopes are presented to the organism, they differ in the ease with which they are assimilated, in the organs in which they accumulate, and the length of time they remain in them. The existence of **critical organs** in which particular substances accumulate is particularly important for α and β emitters, which have an intensely ionizing effect but are damaging only over a short distance. The critical organ for the radioisotope iodine-131 is the thyroid because the thyroid accumulates 600 times the iodine concentration of other tissues in the body. The critical organ for strontium-90, which behaves chemically like calcium, is bone. For manganese-54 the critical organ is the liver.

A great many factors must therefore be taken into account in assessing the possible impact of radioactive effluents on a section of the population:

- the nature of the radionuclide;
- the nature of the emitted radiation;
- its half-life;
- the decay scheme;
- the chemical form in which the radionuclide is encountered;
- the fraction that is assimilated;
- the organ in which it may accumulate and the concentration that may be reached.

The principal source of guidance about radiation exposure is the International Commission on Radiological Protection (ICRP). This body grew out of an international commission set up in 1928 to advise on tolerable levels of X-ray and radiation exposure. There is no evidence that there is a threshold level of radiation exposure below which there are no harmful effects, and when setting tolerable levels of exposure it must be assumed that any increase in radiation levels above the inescapable natural background level, carries some risk of damage. No human activity is without some degree of risk, however, and a degree of risk is acceptable if the benefits of

an activity are great enough. The use of X-rays in medical diagnosis, for example, entails some risk of radiation damage, but given the safeguards that are in use, the risk is an acceptable one.

Legal limits of exposure

Limits to radiation exposure recommended by the ICRP are designed to safeguard those regularly exposed in the course of their work. The overall limit previously prescribed was 50 mSv year^{-1}, and this figure was adopted in an EC directive in 1980 and has become the legal limit in many countries. Particular tissues should not receive more than a fraction of this dose and there are detailed recommendations about the annual exposure to 240 different radionuclides that may be ingested or inhaled.

Occupational exposure to 50 mSv year^{-1} would, on statistical grounds, result in 340 additional deaths per million workers per year. For comparison, in the United Kingdom in 1986–7 there were 300 accidental deaths per million in metal mining, 106 per million in coal mining, 95 per million in metal manufacture, and 92 per million in the construction industry.

In practice, occupational exposure to radiation is well below the legal limit, but for some time it has been recognized that the limit is probably too high and lower limits have been introduced in several countries. In 1987, the British Health and Safety Commission recommended that the circumstances in which any worker was exposed to 10 mSv year^{-1} should be examined. It has recently recommended that any worker exposed to 75 mSv of radiation in any five-year period since the beginning of 1988 should also be subject to review. Although these recommendations do not have the force of law, they have generally had a similar effect.

In 1991, the ICRP revised its recommended limits to 20 mSv year^{-1} and 100 mSv over a five-year period and these limits are likely to be adopted in place of the previous recommendations.

Members of the general public are much

less regularly exposed to radioactivity from man-made sources than those employed in nuclear installations. The ICRP has recommended that for routine exposures, excluding medical and natural sources, the dose should not exceed 1 mSv year^{-1} for lifetime exposure, or 5 mSv year^{-1} for short periods. In the event of accidents, such as that at Chernobyl, control of foodstuffs should be considered if exposure is 5 mSv year^{-1} in the first year, and should certainly have been attempted if the dose is 50 mSv year^{-1}. A whole-body dose of 5 mSv year^{-1} is one-tenth that previously recommended as the limit for occupational exposure. It entails a risk of one to ten radiation-induced deaths from cancer per million of the exposed population. Allowing for the fact that there is wide variation in the levels of exposure among the population, such an upper limit means that the average dose received by members of the public is about 0.5 mSv year^{-1}. This is about the same dose as might be expected from living in a brick house rather than a wooden one.

Critical path analysis

A more stringent safeguard is provided in setting limits to radioactive discharges to the sea by directing attention to the most likely route by which radionuclides may reach the human population, and the dose likely to be received by that section of the population most at risk. The **critical path** approach is exemplified by the way in which it has been applied to discharges from the fuel reprocessing plant at Sellafield.

A large range of radionuclides is included in the liquid effluent, which is discharged from the reprocessing plant by a 2.5 km long pipeline into the Irish Sea. The Irish Sea supports a variety of commercial fisheries and the most critical pathway to man through the consumption of contaminated seafood from the area proved, surprisingly, to be by the seaweed *Porphyra*. *Porphyra* was collected along the coast near Sellafield and shipped to South Wales where, along with *Porphyra* harvested in other parts of Britain, it was processed into laverbread, a delicacy whose consumption is almost entirely limited to South Wales. *Porphyra* accumulates a variety of radionuclides from seawater (Table 7.5), the most important being ruthenium-106. Surveys carried out in South Wales showed that 26 000 people consumed up to 75 g day^{-1} of laverbread regularly, but a small group of 170 adults was identified with individual consumption rates of about 160 g day^{-1}, with a maximum of 388 g day^{-1}.

The critical tissue for ruthenium-106 is the lower large intestine, and for strontium-90, plutonium-239, and americium-241, it is bone. Because of the variety of radionuclides in the *Porphyra*, calculations were made for individual radionuclides, on the most pessimistic assumptions, ensuring that the total exposure did not exceed recommended limits.

The recommended annual dose limit recommended by the ICRP for the lower large intestine was 15 mSv year^{-1}; estimated doses received by the critical subgroup of laverbread eaters in 1962–7 was 4–7 mSv year^{-1}. The recommended genetically acceptable dose was calculated to be 1.3 mSv year^{-1}; the dose resulting from laverbread consumption was 1 μSv year^{-1}. These calculations were based on the very conservative assumption that all *Porphyra* used in the manufacture of laverbread came from the neighbourhood of Sellafield, and ignored the water added during its preparation, which dilutes the radioactive component.

Use of the critical path approach requires constant vigilance to take account of changing circumstances and increasing knowledge of the behaviour of radionuclides. *Porphyra* ceased to be a significant pathway in the mid-1970s when the last two commercial collectors of the seaweed retired and Sellafield *Porphyra* was no longer used in the preparation of laverbread. Locally caught fish then assumed the critical position: the maximum exposure of an individual in the critical population from fish consumption was 34 per cent of the dose limit recommended by the ICRP. Caesium-137 contributes substantially to radiation exposure from eating fish caught around Sellafield, and caesium discharges rose substantially in 1974–5 as a result of changed practices at the reprocessing plant. In 1976, the critical group among members of the public received 25 per cent of the ICRP recommended dose limit of caesium-137 from this source. In subsequent years, this figure was reduced to about 10 per cent of the ICRP limit. Further reassessment of exposures was necessary in 1982 when a new survey by the Ministry of Agriculture, Fisheries, and Food revealed that the estimated consumption of molluscs by the critical group of seafood consumers had increased threefold. Their intake of plutonium and americium had increased by a similar factor. It had also been discovered that the uptake of plutonium from food by the human gut is five times higher than was previously thought. Thus, the calculated dose of plutonium received by members of the critical group must be increased by a factor of 15, bringing their total exposure from 24 per cent to 39 per cent of the ICRP annual dose limit, of which actinides (plutonium and americium)

contribute 26 per cent rather than the 4 per cent previously estimated. Steps were then taken to reduce the discharges of α radioactivity accordingly.

Critical path analysis is used to determine discharge standards in a number of countries and this accounts for the great variation in the limits set for each installation, which must take account of local circumstances and the critical pathways of exposure of the human population to radioactivity. The nuclear power station at Lake Trawsfynydd in Wales discharged 5.18 TBq in 1975 and individuals in the critical group consuming fish from the lake are estimated to have received 8 per cent of the ICRP limit. The reprocessing plant at Sellafield, on the other hand, discharged 10 000 TBq in that year, when maximum exposure of a member of the critical group of seafood consumers was estimated to be 34 per cent of the ICRP limit, only four times as much despite the enormously greater discharge.

DREDGINGS, SOLIDS, PLASTICS, AND HEAT

DREDGINGS

Very large quantities of sediment are dredged from the seabed for various uses on land or to prevent the silting of shipping channels, and some or all of it is returned to the seabed at a different site from where it originated. In addition, some inert industrial solids are dumped at sea where they have similar effects on benthic communities to dumped dredged material.

Sediment extraction

Sand and gravel are increasingly mined from the seabed in many parts of the world, and are used in land reclamation, replenishing eroded beaches, building artificial islands or break-waters, and for use as construction aggregates, mainly in concrete production. In south-east Asia, tin ore is mined by dredging sediments in shallow coastal waters. In Scandinavia, glacial boulders (erratics) weighing 0.2–10 t, lie on the surface of the seabed and are extracted for the construction of breakwaters, and in coastal protection by the use of special vessels called 'stone fishers' fitted with large claw-like grabs.

Extraction of sand and gravel is either by anchor dredging or trailer dredging (Fig. 8.1). In the former case the dredger is anchored over the deposit and mines it by forward suction through a pipe, leaving pits on the seabed up to 20 m deep and 75 m in diameter. Trailer dredgers extract the deposit by backward suction through one or two pipes while under-way, leaving shallow furrows 20–30 cm deep and up to 2 m broad. The aggregate is piped

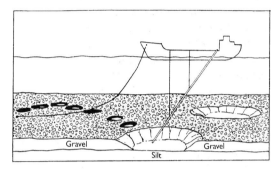

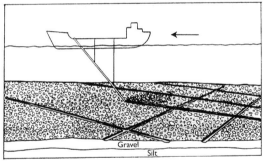

Fig. 8.1 Two methods of dredging marine aggregate: (*upper*) anchor dredging, (*lower*) trailer dredging.

into the ship's hopper, displacing water which carries mud and silt back into the sea, forming a turbidity plume. In some cases, the aggregate is screened on board and excess sand and pebbles are returned to the seabed to maintain a specific sand-pebble ratio in the cargo.

Tin ore is mined by suction or bucket dredgers; the dense ore is extracted by settlement on board the dredger and large quantities of the less dense sediment are then returned to the sea.

A direct effect of sand and gravel extraction

is the loss of the spawning and nursery habitat for fish such as herring (*Clupea harengus*) or, in the Baltic, coregonids (whitefish and vendace), which are fished intensively. Sand eels (*Ammodytes*) form the basis of industrial fisheries but are probably more important as food for other fish and sea-birds; they live permanently in the sand or gravel beds and lay their eggs on sand, where they are sensitive to smothering. Although there are fears that sand and gravel extraction threatens these fisheries, there has been little evidence of damage so far. Studies in a sand extraction area in the Gulf of Finland failed to show any effect on herring spawning in nearby spawning rounds, but local catches of herring were noticeably smaller, perhaps because of the disturbance caused by the noise of sand extraction.

Recovery after dredging

The rate at which the topography of the seabed is restored after dredging depends upon bottom currents and, except in areas of shifting sands, is generally slow. Indeed, after the experimental removal of sand and gravel from a rocky area of Seine Bay, on the French coast, there was no sign of the restoration of the previous cover and a typical hard bottom fauna developed.

The large pits left after anchor dredging of sand in the Dutch Wadden Sea filled in one year in tidal channels, but in other areas required 5–10 years to fill, and those on tidal flats required more than 15 years. Furrows in gravel beds caused by trailer dredging have a more widespread impact but are much shallower and recovery is generally quicker, although in one test site off the east coast of England furrows were still visible four years after dredging. The time for recovery from sand extraction in the North Sea averages three years.

The recovery of the fauna of dredged areas depends in part on how the dredging is carried out, since this influences the availability of recolonizing organisms as well as the suitability of the affected site for recolonization.

Following maintenance dredging at one site

in a shipping channel at Coos Bay, Oregon, it was found that the sediment had not been removed uniformly from the channel but furrows had been created, separating undisturbed hummocks of sediments containing adult populations of organisms capable of repopulating the dredged areas between; these areas acquired a predredging fauna twice as quickly as other areas where dredging had been more thorough. The benthic community of this shipping channel readjusted to predredging conditions in only 28 days and was evidently adapted to a naturally stressful environment. In addition to the impact of periodic dredging, the community was subject to disturbance by tidal scour, prop-wash from passing marine traffic, as well as the discharge of industrial, domestic, and shipping wastes.

Where there is greater dredging impact, however, readjustment of the fauna usually takes a different course. Particularly where deep pits are formed by anchor dredging, the speed of bottom currents is reduced locally and this results in the infill being formed of fine sediment and it may also be subject to periodic or long-term deoxygenation which inhibits the establishment of a new fauna. Even without this complication, it is commonly found that infill sediments differ from the original substratum and have their own characteristic fauna. Dredged areas may therefore develop a mosaic of habitats.

Impact of sedimentation

Sediment returned to the seabed as washout from the dredger, or by direct dumping, has an impact on the benthos of the receiving area, depending on the nature and sensitivity of the local benthic communities and the rate of sedimentation. Mobile epifauna and active burrowers in the area where sediments are deposited are least likely to be affected by smothering, although the actual impact will depend on the taxa and rate of deposition of sediment.

One study carried out in the fjord at Uddevalla on the west coast of Sweden (Fig. 8.2) where dredging was going on, was in an area where surveys of the benthic fauna had

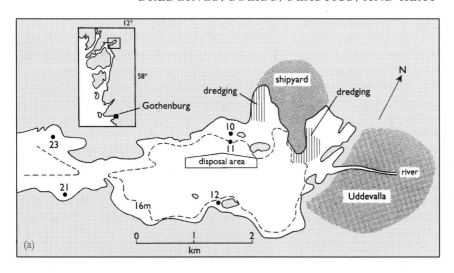

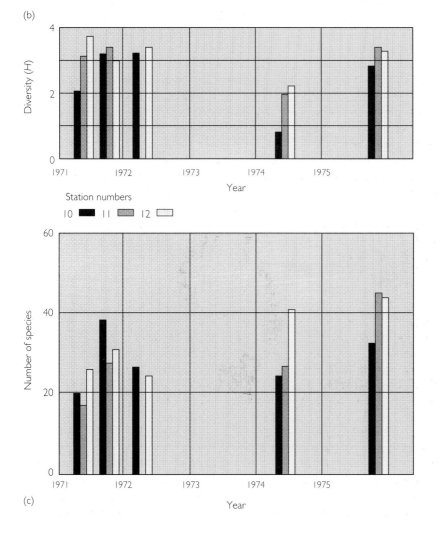

Fig. 8.2 (*a*) Dredged and disposal areas in the Byfjord, western Sweden. (*b*) Diversity (Shannon–Wiener index), and (*c*) number of species at stations 10, 11 and 12 before and after dredging; station 12 is the control site. (*Elsevier Science*)

previously been made. Immediately after dredging, there was a loss of diversity at stations near the operations because of a failure of recruitment of several bivalves, probably because of increased sediment in the water, but the position was restored a year later.

In another study following dredging of a sandbank off the east coast of Queensland, Australia, sediment deposition from the turbidity plume actually resulted in an increase in species diversity and abundance (Fig. 8.3), but this appears to be an unusual experience.

On the other hand, heavy sedimentation downstream of tin mining operations on the western coast of Thailand has had a serious and prolonged impact on mangrove forests and coral reefs. Following mining, the soil in mangrove areas suffers a drastic reduction of organic matter and nutrients, making it unsuitable for recolonization by the plants, and restoration requires 20 years. Coral reefs suffer as a result of smothering, although regeneration is possible if water turbulence removes the sediment during monsoon periods. Coastal tin dredging continues, but on a reduced scale following the collapse of the tin market in the mid-1980s.

Contaminated dredgings

Ports, harbours, rivers, and approach channels often need regular dredging to keep them open for shipping. The dredged material is often anoxic and, particularly if it comes from harbours or industrialized estuaries, is usually contaminated with metals, pesticides, and persistent oils, the legacy of a century or more of discharges into the rivers (Table 8.1). The

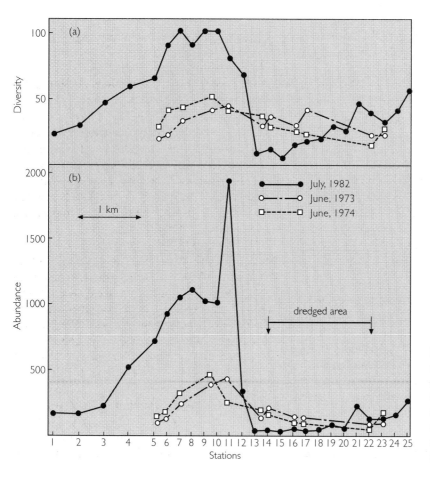

Fig. 8.3 Abundance (number of individuals) and diversity (number of species) in transects across a shipping channel before and after dredging. (*Springer-Verlag*)

Table 8.1 Quantities of metals (t) in dredging spoil dumped in the North Sea and English Channel in 1990

	Total ($\times 10^3$ t)	Cd	Hg	Cu	Pb	Zn	Cr	Ni
Belgium	23 447	47.5	7.21	339	940	2338	764	448
Netherlands	20 225	6.0	2.40	165	325	1180	410	130
France	10 528	2.73	0.70	64	177	550	195	53
UK	7305	1.70	1.42	153	301	582	115	85
Total*	64 787	58.4	11.9	835	1765	4750	1605	718

*Including smaller amounts from Denmark and Germany.

contaminants are transferred to the dumping grounds.

The nature of the contamination varies widely. Dredged spoil from the docks of the Manchester Ship Canal contains 20.7 ppm mercury and 5080 ppm lead, while spoil from the Tees estuary (also with long-standing industrial inputs) contains 7 ppm mercury and 3000 ppm zinc, but only 320–460 ppm lead. Spoil from Swansea in South Wales contains a high concentration of cadmium (18 ppm), perhaps from natural sources.

If the dumped spoil stays in piles on the seabed, it may remain anoxic and the contaminants will then be biologically unavailable and so do no harm. However, if the dredging spoil becomes dispersed and oxygenated, metals may change their chemical species (see Chapter 5), cease to be adsorbed on to silt particles, and then enter food chains.

Because of this danger, severe restrictions have been introduced on the disposal of contaminated dredgings in a number of sea areas, and alternative disposal methods have been used.

• Dredgings from the inner harbour at Rotterdam and from the River Elbe at Hamburg are no longer dumped in the North Sea but are deposited on artificial islands. These have a limited capacity but it is hoped that the Rhine and Elbe, the sources of most of the contaminants, will have been sufficiently cleaned for sea dumping to be resumed in due course.

• In Hong Kong and Florida, contaminated dredgings have been dumped in a deep exca-vation offshore and then capped with uncontaminated sediment.

• For some years, contaminated dredgings from the Thames estuary were dumped on marshes bordering the estuary and eventually turned into pasture.

INDUSTRIAL WASTES

Mine tailings

During mining, large quantities of non-target material are excavated in the construction of underground tunnels and galleries, as well as being extracted along with the target material. This unwanted material is separated from the target material in sorting plants on the surface and is frequently dumped near the mine in large spoil heaps which significantly change the local landscape. For coastal mines, sea disposal is an attractive alternative.

In several parts of the world, inert but bulky mine waste is discharged to the sea without damaging coastal or inshore faunas by a process known as **deep submarine tailings disposal**. Mine tailings are piped as a slurry to the edge of a submarine canyon. The waste is discharged at such a density and speed on to the slope that a 'turbidity' or 'density' current is produced. This moves the waste with little dispersion or upwelling down the slope into deep water (often more than 1000 m). This system has been used for many years in British Columbia by a copper mine where the discharge is into an adjacent deep fjord. The system is also used in mines in Papua New Guinea and the Philippines where the drop-off at a coral reef edge is used.

As with shallow water dumping, the waste smothers the benthos, but the deep water fauna is sparser than that in shallow water and local fisheries are not dependent upon it. In these circumstances, marine disposal is regarded as environmentally preferable to dumping the mine tailings on land.

For many years, colliery waste was dumped by barge in 30–50 m of water off the north-east coast of England, or, in some cases, deposited on the shore by conveyor belt from the pit head. This coastal dumping continued until the coal mines closed in 1992. The material dumped is chemically similar to the natural sediments of the region, but the particle size distribution is different and the continual addition of material smothered the seabed so that the subtidal areas suffered impoverishment of the fauna.

Damaging effects were most pronounced where colliery waste had been deposited on the shore by conveyor belt. There has been little dispersion and beach levels have been raised several metres with the total loss of the natural shore fauna. On the other hand, it has provided considerable protection from coastal erosion.

Experience of former dumping grounds for mine tailings off the English and Canadian coasts has shown that a productive ecosystem is usually restored within a year or two after dumping stops. However, an area that had received tailings from a copper mine at Britannia Beach in western Canada for some 75 years, still remained impoverished in a number of taxa 12 years after the mine closed. This may have been owing to the extensive contamination of the shallow water sediments with copper.

China clay waste

China clay deposits near St Austell in Cornwall, south-west England (Fig. 8.4), have been used since the eighteenth century as a source of fine clay for the manufacture of porcelain. Extraction of the clay produces a waste consisting of fine sand, kaolin, and flakes of mica,

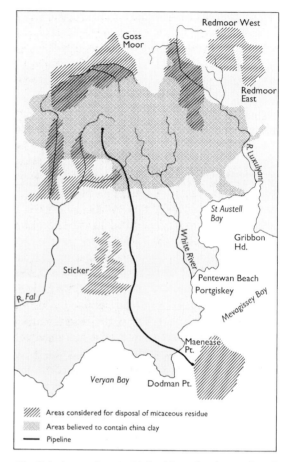

Fig. 8.4 Areas from which china clay is produced near St Austell, Cornwall, and the discharge routes for china clay waste. (*Elsevier Science*)

which has been discharged into the rivers Fal, Luxulyan, and St Austell, although the Fal ceased to be used for this purpose many years ago and discharges to the other rivers are now much reduced.

Most of the river-borne waste is deposited in St Austell Bay and Mevagissy Bay (Fig. 8.5(a)). The substratum in the area is naturally one of coarse sand, shell, pebble, and rock, but about three-fifths of the seabed is now covered, sometimes to a considerable depth, with micaceous clay waste and this has caused faunistic changes. Some bivalves, such as *Clausinella* (= *Venus*) *fasciata* (Fig. 8.5(b)) and *Dosinia exoleta*, are intolerant of the clay waste, but there is a characteristic fauna of the china clay waste, including the polychaete

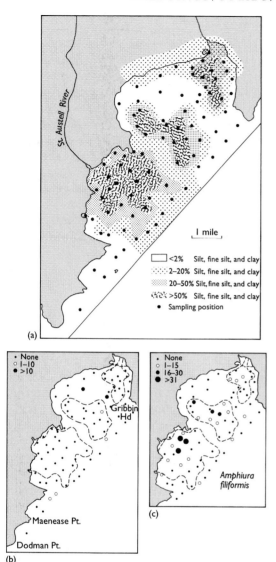

(a)

(b)

(c)

Fig. 8.5 St Austell and Mevagissy Bays, Cornwall: distribution of (*a*) china clay waste fractions in bottom sediments, (*b*) the bivalve *Clausinella* (=*Venus*) *fasciata* intolerant of clay, and (*c*) the brittle star *Amphiura filiformis*, clay tolerant.

Melinna palmata, the bivalves *Fabulina* (=*Tellina*) *fabula* and *Phaxas* (=*Cultellus*) *pellucidus*, the heart urchin *Echinocardium cordatum*, the holothurian *Labidoplax digitata*, and the ophiuroid *Amphiura filiformis*. Plaice (*Pleuronectes platessa*), dab (*Limanda limanda*), and sole (*Solea solea*) feed on this specialized fauna.

Fly ash

The very fine, powdered fly ash produced in coal- and oil-fired power stations is composed mainly of silicon dioxide with oxides of aluminium and iron, together with trace concentrations of metals. The greater part of it is extracted from the exhaust fumes in the smoke stacks by air pollution control equipment. The rest is the very fine powdered ash produced in the furnaces of coal-fired power stations.

There is some use of fly ash in the construction industry, but supply greatly exceeds demand. The surplus is generally dumped on land, but some is dumped at sea.

Between 1951 and 1992, when the practice ceased, up to 600 000–700 000 t year^{-1} of fly ash were dumped at licensed grounds off the north-east coast of England. The particle size of the ash is 2–200 μm and is similar to that of fine silt, but because of its homogeneity, it offers low niche diversity. At the dumping ground some species, such as the gastropod *Turritella* and the foraminiferan *Astrorhiza*, which were common when the dumping started, became rare or were eliminated.

Unlike the situation in St Austell Bay with an input of china clay waste, where there has been a compensating development of a rich benthic soft-bottom fauna, the main dumping ground for fly ash shows extreme impoverishment of the fauna, presumably because of the daily disturbance of dumping by barge. As in St Austell Bay, the fine material has sifted into rock crevices to the detriment of a lobster fishery.

ARTIFICIAL REEFS

Coastal protection

Artificial reefs are constructed for a variety of purposes, but the commonest is protection of the coast from wave action. Dredging spoil is commonly used for this purpose and 4000 such breakwaters have been constructed in Japan, alone, in the last 20 years.

Fisheries enhancement

It has long been known that wrecks provide a haven for fish and in the Gulf of Mexico offshore oil platforms shelter a variety of exotic fish and have been valued by sport fishermen for many years. Appropriate scrap material may be deliberately dumped at sea to enhance local fisheries.

- The most effective materials for artificial reefs are old ships, and in a number of designated sea areas 'graveyards' for obsolete shipping have been established in Australia and elsewhere. The vessels have to be cleaned and all floatables removed before they can be disposed of in this way.

- Redundant oil platforms have been used as artificial reefs and have attracted fishermen off the coast of southern California since the 1960s. In the Gulf of Mexico rigs have been purposely collapsed to serve as artificial reefs; this may involve moving some to deeper water to avoid interference with shipping.

- Old car tyres, baled together, have been used very successfully, particularly in the USA, to construct fish reefs. The bales of tyres are stable and provide a suitable habitat for fish and a large surface area for colonizing organisms.

- Trials using concrete blocks made with surplus fly ash from coal- or oil-fired power stations have been carried out in Japan, New York, Florida, and southern England. They proved to be suitable material for fishery enhancement, but economic considerations have prevented the use of fly ash on a serious scale for this purpose.

- Other materials, such as concrete waste, pipes, and a variety of bulky solids, have been used successfully. Old car bodies corrode too quickly to be effective.

Opposition to the disposal of the redundant oil platform *Brent Spar* in the deep North Atlantic in 1995, evidently did not take account of these practices.

Shipwreck

The seabed is littered with vessels lost through shipwreck in areas hazardous to shipping, or, in wartime, by military action. Presumably the fuel oil and cargoes, which may have included noxious material, had some deleterious impact on the local marine environment, but if so it has rarely been recorded and there is no evidence that any harm done has been of long duration.

While the wrecks may present a hazard to fishing because of the risk of snagging nets, they provide a sheltered environment for fish, are attractive to recreational divers, and in many cases are protected as archaeological sites.

LITTER AND PLASTICS

Litter

The seas in all parts of the world are littered with man-made debris, most of it plastics, which are practically indestructible. In recent studies it was found that in the eastern Mediterranean 70% of bottom trawls were fouled with debris, in Alaska 57 %, and even in the Bering Sea over 40% of bottom trawls contained litter. One study in Indonesia found that over half the gill net expeditions by artisanal fishermen had debris fouling the nets. Most of the litter is eventually stranded on the shore line; as much as 1130 kg km^{-1} was found on one shore in Georgia, USA.

Occasionally the litter is known to be derived from onshore sources, such as waste tips on the coast, but most comes from shipping and high concentrations of plastic debris are found near busy shipping lanes and fishing areas. About 2 kg of rubbish per person per day is generated on ships and for a cruise liner in the Caribbean, the amount of garbage to be disposed of can easily overwhelm the facilities of the ports it visits.

Annex 5 of the International Convention for the Prevention of Pollution from Ships (MARPOL) came into effect in 1988 and bans the disposal of waste other than food to the sea by ships, but it is widely disregarded and an estimated 6.5 million t year^{-1} of plastics are discarded by ships, most of it within 400 km of land.

Most of the litter at sea or on the coast is primary or secondary packaging, commonly plastic bags, cups and bottles, tampon applicators, and pieces of polystyrene. In addition, on the North American coast, glass bottles, broken glass, and metal beverage cans, and in southeast Asia discarded plastic footwear, are abundant in beach litter. North Pacific coasts are littered with logs, some derived from the lumber industry, but most of natural origin.

Plastics are inert to marine organisms and while floating at sea may acquire a fauna of encrusting organisms such as algae, barnacles, hydroids, and tunicates. Larger animals ingest plastic objects: large fish and 59 per cent of beached turtles have been found with plastic cups and other objects lodged in their digestive tract. A more serious threat is entanglement and there are numerous records from places as far apart as Alaska, South Georgia, Tasmania, South Africa, and Nova Scotia of young seals and sealions becoming tangled with fishing gear, often forming a collar around the neck causing deep lacerations.

The cause is commonly discarded trawl net, rope, plastic straps, monofilament long line, or buoy ropes, but among the more bizarre reports is one of a seal trapped in a wind sock and another of a seal wearing a floatation vest, complete with whistle and CO_2 cylinder. Dogfish, porbeagle sharks (*Lamna nasus*), and yellowfin tuna (*Thunnus albacares*) have at various times been observed entangled in plastic wrapping bands. Common dolphins (*Delphinus delphis*), bottlenose dolphins (*Tursiops truncatus*), and two southern right whales (*Eubalena australis*) have been reported entangled in fishing gear in Australian waters.

Sea-birds, most often gulls, but also including gannets (*Sula bassana*), Brunnich's guillemot (thick-billed murre, *Uria 'lomvia*), ducks, geese, ospreys (*Pandion haliaetus*), and various wading birds, also become entangled and killed by trawl net, six-pack yokes, and other debris. Stranded debris can also be a threat to land animals: on the Nova Scotia coast, a

horse was found with both hind legs entangled by a bundle of plastic strapping.

Plastic particles

Small particles of polyethylene, polypropylene, and sometimes polystyrene, 3–4 mm in diameter, are widespread in all oceans. Their presence has been reported on beaches and in surface waters near centres of industry or major shipping routes since the early 1960s, but they have also been found in the south Atlantic and south Pacific several thousand miles from any known source, and it must be assumed that their distribution is worldwide.

These small particles include weathered fragments of larger plastic articles, but a high proportion are plastic pellets which are 'feedstock' for the plastics industry and reach the sea through accidental spillages at ports or factories close to rivers. It is not certain if they are harmful to marine organisms, but they have been found in the stomach contents of flying squid (*Ommastrephes bartrami*) and a variety of turtles. They are ingested by fish, and there is evidence that white, but not blue or transparent plastic particles are selectively ingested by larval and adult fish, which have been found to contain only white, opaque polystyrene particles.

They are also ingested by surface-feeding sea-birds such as shearwaters, petrels, prions, and phalaropes and some planktivorous diving birds such as auklets. Prions (*Pachyptila*) are small petrels from the southern oceans, which have a sieving arrangement on the lower bill and feed on plankton. They commonly have numerous plastic particles packed in the gizzard. In Alaska and the Aleutian Islands, plastic ingestion has been reported in 37 species, and the incidence of plastic ingestion had increased markedly between 1969–77 and 1988–90. Large carnivores, such as great skuas (*Catharacta skua*), acquire plastic particles from their prey and regurgitate them along with other indigestible material in their pellets. Adult birds may pass plastic fragments on to their chicks when feeding them by regurgitation.

Although one can speculate that an accumulation of plastic particles in the gut may be harmful, the only evidence of this is a negative correlation between body fat and the number of particles in migrating red phalaropes (*Phalaropus fulicarius*) on the Californian coast. If ingested plastic particles are responsible for impairing feeding, this would obviously be detrimental to birds facing a long migration.

Drift nets

Drift nets are long, vertical curtains of net used to catch fish that shoal near the surface. They are supported by floats at the top of the curtain and the lower edge is weighted. If they break free, they continue to catch fish automatically and continuously ('**ghost fishing**') until they are washed ashore or rot. This is not a new problem, but it has been greatly increased by the introduction of nylon monofilament gill nets.

Such nets are virtually indestructible and are very efficient at catching fish, perhaps because the fish cannot detect them. It is claimed that the catch of bass (*Bicentrachus labrax*) on the coast of south-west England increased from 100 kg to 3000 kg per fishing vessel per day after the use of monofilament gill nets was introduced.

Unregulated use of drift nets resulted in the over-exploitation of a number of fish stocks, particularly in the Pacific. The Taiwanese shark fishery, mainly for black tip shark (*Carchorhinus tilstoni*) and sorrah shark (*C. sorrah*) in the Arafara Sea, north of Australia, peaked at over 20 000 t in 1978 followed by a fall in the catch per unit effort, evidence that the stock was being over-exploited. Similarly, the Japanese fishery for albacore tuna (*Thunnus alalunga*) in the Tasman Sea rose from 2000 t in the early 1980s to an estimated 25 000–49 000 t in 1989–90, but back to 7500 t in the following season. As a result of these and similar experiences, the Canadian, New Zealand, Australian, and South African fishery authorities found it necessary to ban or closely regulate the use of drift nets in their waters. A total ban on their use throughout the world was introduced by a United

Nations' resolution and came into effect at the end of 1992.

Drift nets, especially monofilament gill nets, do not catch fish alone; a large number of birds and sea mammals are trapped by them and drown. Since the nets are commonly 6 m deep and may be 60 km long, their capacity to inflict damage if they break free is very great indeed. Reported casualties include northern fur seals (*Callorhinus ursinus*), dall porpoises (*Phocoenoides dalli*), sea otters (*Enhydris lutris*), and, on the south-east African coast, Cape fur seals (*Arctocephalus pusillus*). It has been reported that 130 000 small cetaceans are caught in nets each year, but the actual number may be much higher.

Sea-birds trapped by gill nets include the laysan albatross (*Diomedea immortabilis*), fulmars (*Fulmar glacialis*), shearwater (*Puffinus griseus* and *P. tenuirostris*), and tufted puffins (*Lunda cirrhata*). The gregarious, diving auks are particularly vulnerable. The Danish salmon drift net fishery in the north Atlantic is estimated to have killed 500 000 ± 250 000 sea-birds, mostly Brunnich's guillemot (*Uria lomvia*), each year during the period 1965–75 when the fishery was exploited. The Japanese salmon fishery in the north Pacific and Bering Sea took a similar annual toll: between 214 500 and 715 000 during 1952–75, and 350 000–450 000 in 1975–78. Such losses are an order of magnitude greater than the estimated losses of sea-birds from oil pollution in these areas.

MUNITIONS

Many countries have disposed of defective, obsolete, or surplus munitions at sea. Accidental losses in wartime, and smaller losses of explosives and target ammunition seaward of firing practice ranges also remain on the seabed. The wreck of the ammunition ship *Richard Montgomery*, containing 5000 t of aerial fragmentation bombs and 1100 t high explosives lying in the Thames estuary only 4.8 km from the Isle of Grain oil terminals, is but one example of a continuing major hazard.

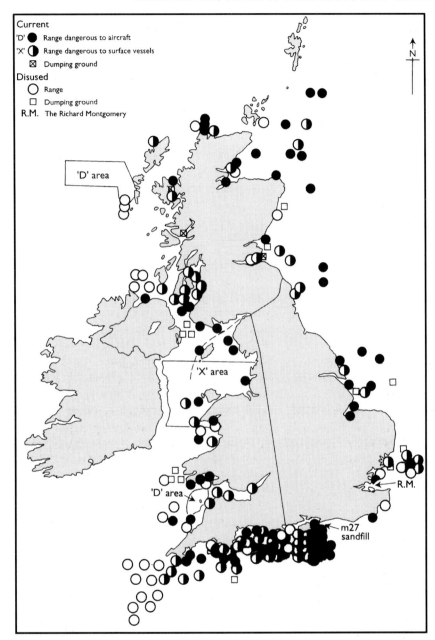

Fig. 8.6 Main sites of munitions inputs to British coastal waters. (*Elsevier Science*)

In the post-war period until 1957, Britain scuttled 24 ships packed with 137 000 t of chemical weapons at deep water sites in the Atlantic, but many weapons have been disposed of in much shallower water. Dumping grounds and known dangers to shipping are indicated on charts. Figure 8.6 shows the main sites of munitions inputs to British waters, and the coastal waters of other countries harbour similar hazards, although the information about them has not been analysed.

Munitions disposed of in shallow water may be shifted from their original dumping site by water movements; furthermore, fishing practices have changed and deeper waters are trawled than when the dumping took place. A considerable number of munitions are recovered in the course of fishing and dredging

Table 8.2 Types of munitions recovered and the proportion that had to be destroyed *in situ* in the Portsmouth and Medway Command Area (PMCA), 1969–77

Class of munitions	Number	% destroyed
Bombs	422	81.5
Land service mortars and grenades	141	37.5
Mines and torpedoes	51	26.0
Anti-submarine weapons	11	23.4
Shells	1303	21.1
Rockets and propulsion units	23	13.6
Pyrotechnics	46	6.9
Total	1997	24.4*

*Percentage of all finds destroyed.

operations, and a great variety come ashore, often in a dangerous condition (Table 8.2).

There are periodic reports of human injuries from chemical and incendiary devices, in British waters and the southern Baltic Sea, but munitions in the sea have no material impact on living resources.

Marine disposal of unwanted munitions is now regulated and deliberate dumping in coastal waters is not authorized except in an emergency. The inputs over many earlier years, however, remain a hazard to users of the sea.

HEAT

Cooling water and often other industrial effluents are discharged at a higher temperature than that of the receiving waters. By far the greatest amount of heat discharged to the sea is in cooling water from coastal power stations. About 20 million m³ of cooling water, 12 °C above the ambient sea temperature, are discharged for 1000 MW of electricity generated by oil- or coal-fired power stations. Nuclear power stations are a little less efficient and the cooling water from them is about 15 °C above ambient.

In the tropics there is little fluctuation in the power demand or the sea temperature, but in temperate regions, the peak load, and so the greatest discharge of cooling water, is in winter when the sea temperature is low and falling. In subtropical areas, particularly those of North America, the peak load is in summer because of the extensive use of air conditioning. The maximum discharge of hot water is thus at a time when sea temperatures are near their maximum which may be as high as 30–35 °C. Not only is the added heat dissipated more slowly than in cold seas, but the sea temperature is already near the thermal death point for many organisms and, with these additions of heat, may easily exceed it.

Cooling of the heated discharge is almost entirely by mixing with the receiving water. The area affected is limited to the plume of hot water and its immediate surroundings, although the direction taken by the plume may change with changing tidal currents and so the total area under its influence is greater than appears at first sight. Even so, the outfall from the Diablo Canyon nuclear power station on the Californian coast affects only 0.7 ha; the 250 MW Bradwell nuclear power station in Essex, discharging 2 million m³ day⁻¹ into the enclosed Blackwater estuary, raises the temperature of the surface water by only 0.2–1.7 °C. A succession of power stations using and reusing the same water along the tidal Thames have a total capacity of 8000 MW, but do not cause a build-up of water temperature. In contrast, in subtropical

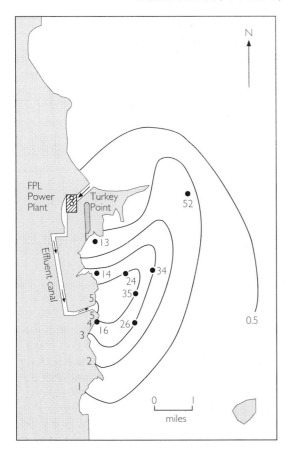

N

FPL
Power
Plant

Turkey
Point

Effluent canal

• 52

• 13

• 14
  24
  35 •
• 34

5
5
4
3

• 16   26 •

0.5

2

1

0   1
       miles

Fig. 8.7 Increase in surface sea temperature resulting from the discharge of cooling water from power stations at Turkey Point, Florida. (*Elsevier Science*)

waters, much larger areas may be affected: at Turkey Point, on the Atlantic coast of Florida, nearly 40 ha are affected by the cooling water discharge from a power station complex (Fig. 8.7).

It is difficult to separate the effects of hot water from other features of the discharge. Cooling water is usually treated, at least intermittently, with chlorine to discourage the settlement of organisms in the heat-exchange system. The scour of the seabed caused by the water flow in the plume is likely to change the nature of a soft substratum and so influence the fauna. Metals are leached from the cooling system and the disappearance of abalones (*Haliotis*) from the neighbourhood of the Diablo Canyon power station is attributed

to copper in the cooling water, not to the elevated temperature.

The Turkey Point power station complex discharges cooling water in a situation where quiet, shallow soft-bottom areas are dominated by turtle grass (*Thalassia*). This marine grass forms the basis of a specialized community supporting a characteristic fauna. The temperature of the sea around the discharge is 30–35 °C in summer and a further increase of 5 °C or more is damaging to *Thalassia*. About 9.3 ha of *Thalassia* has been destroyed by this outfall and a further 30 ha shows reduced growth.

Here and elsewhere it has been found that tropical marine animals are generally unable to withstand a temperature increase of more than 2–3 °C, and most sponges, molluscs, and crustaceans are eliminated at temperatures above 37 °C, although intertidal species appear to be more tolerant of high temperatures than benthic species. It is not surprising that there is a reduction in benthic diversity in the area of cooling water discharges from power stations on the Florida coast and in Guanilla Bay, Puerto Rico; the temperature of the discharge reaches 40–45 °C in summer.

In temperate regions hot water discharges, far from being damaging, enhance the growth of bivalves and fish and advantage of this is taken for mariculture of a number of organisms. In enclosed areas receiving heated effluent, several species breed continuously and this has been observed in the ascidians *Ciona* and *Ascidiella*, and the amphipod *Corophium*; others have an unusually long breeding season as in *Balanus amphitrite* and *Elminius modestus*, both exotic species to Europe. Zooplankton blooms earlier in heated water: near the Hunterston nuclear power station in the Firth of Clyde, the copepod *Asellopsis intermedia* blooms two months earlier than elsewhere in the region. Unfortunately, the phytoplankton on which the copepod depends, blooms in response to increasing day length, not sea temperature, so that the copepod does not benefit from its earlier appearance.

Strong-swimming fish are able to avoid the

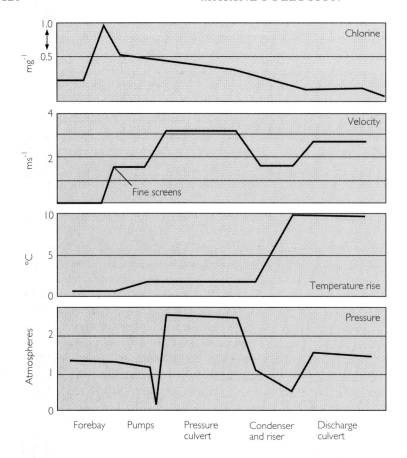

Fig. 8.8 Sudden changes in conditions experienced by organisms entrained in the cooling water flow at a power station.

intake of seawater cooling systems, but smaller fish and plankton are entrained in the flow of water and organisms small enough to pass through the screens are swept through the cooling system. They may be exposed to sudden injections of chlorine, sudden and large temperature changes, and to mechanical buffeting (Fig. 8.8). There is conflicting evidence about the impact of entrainment in the cooling water circuit. The most serious insult to entrained organisms appears to be the mechanical stresses imposed on them; small organisms, such as polychaete larvae, appear to be little affected, but very large numbers of fish eggs and larvae may be killed.

9

THE STATE OF SOME SEAS

In this chapter, a brief view is taken of five seas: the North Sea, Mediterranean, Baltic, Caribbean, and Caspian. All receive a considerable volume of effluents but because of differences in climate, physiography, and hydrographic regimes, the problems in each differ.

THE NORTH SEA

The total area of the North Sea is 575 000 km². The southern end is constricted at the Straits of Dover, but there is no geographical northern boundary, although for practical purposes this can be regarded as a line from Shetland to the Norwegian coast near Bergen. The south and south-eastern parts are shallow, with depths mostly less than 50 m. The northern part is 45–120 m deep, with deeper water off the Norwegian coast. It is not a homogeneous body of water and a number of areas can be distinguished between which water exchange is relatively slow (Fig. 9.1). Flushing time in these areas varies between 46 and 333 days, although surface water to a depth of 10 m is exchanged more rapidly. Overall, there is an excess of precipitation over evaporation, but in winter the lee-effect of the British Isles produces a net loss of water by evaporation in the western and south-western parts of the North Sea. During summer, all parts of the North Sea receive an excess of water by precipitation, and, with surface water becoming less saline, stratification of warm, less dense water over the deeper water is encouraged.

The North Sea is used as intensively, for as

great a variety of purposes, as any sea area in the world. The Straits of Dover and the southern North Sea are among the most heavily trafficked sea lanes in the world; gravel and sand for the construction industry is dredged in a number of areas; gas and oil are extracted in the central and northern North Sea. A largely urban and industrialized population of 31 million lives around the North Sea and its estuaries, and to these are added an influx of summer visitors, particularly to the sandy

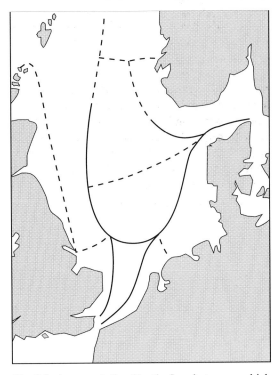

Fig. 9.1 Areas of the North Sea between which there is little water exchange. Broken lines indicate only partial separation of water bodies.

beaches. The North Sea receives the wastes from this population, together with those carried in a number of major rivers: the Rhine, Elbe, Weser, Scheldt, Ems, Thames, Trent, Tees, and Tyne. It also provides the site of highly productive fisheries which are intensively exploited.

Eutrophication

In recent years there has been growing concern about eutrophication in the North Sea. There have been various alarm signals, chiefly in continental coastal waters. Between 1962 and 1985, the phytoplankton biomass at Heligoland in the German Bight increased fourfold and plankton community structure changed. This was correlated with an increase in plant nutrient levels. A further sign of eutrophication was the deoxygenation of bottom waters accompanied by the disappearance of fish and death of the benthic fauna which was observed in the German Bight in three consecutive years in the early 1980s (see Fig. 3.15, p. 35). In 1988 an unprecedented bloom of the flagellate *Chrysochromulina polylepis* spread from Danish waters (Fig. 9.2) around southern Norway, causing great damage to salmon farms.

The areas most seriously affected are the German Bight and Danish coastal waters. These are shallow and often develop a thermocline in summer. The outflow from the rivers

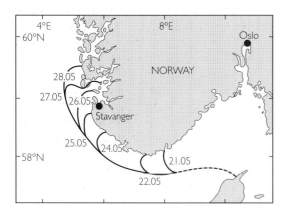

Fig. 9.2 Spread of the bloom of *Chrysocromulina* along the Norwegian coast in May 1988. Dates indicate the position of the 'front' on successive days.

Rhine and Elbe moves northwards up the coast of Denmark and does not mix appreciably with the central part of the North Sea. This water carries a large burden of plant nutrients derived from agricultural runoff, and variations in phytoplankton biomass are correlated with the flow rates of the rivers.

Measures to reduce the input of nutrients were introduced after the 1985 North Sea Conference. At that time, 103 000 t year^{-1} of phosphorus and 1 506 000 t year^{-1} of nitrogen entered the North Sea from river inputs, atmospheric fallout, direct discharges, and the dumping of sewage sludge. By 1995, phosphorus inputs had been halved by all North Sea states except France (25 per cent reduction). Nitrogen inputs had been reduced by 25 per cent, failing to achieve the objective of a 50 per cent reduction. Discharges of untreated sewage from coastal towns has been reduced to a very low level or eliminated altogether in most countries, although 13 per cent of sewage discharges from France, and 50 per cent from Belgium are still untreated. Sewage sludge dumping at sea will have ceased altogether by 1998. Intensive animal farming in Denmark and the Netherlands has had to be limited because of the problems of disposing of the large amount of waste generated.

Conservative pollutants

Since the 1970s there has been a concerted effort to reduce the input of conservative pollutants to the North Sea. This has been achieved by changed practices, such as the adoption of mercury-free technology in the chlor-alkali industry, or the dumping of highly contaminated dredgings from Rotterdam and Hamburg harbours on artificial islands instead of at sea, as well as by stricter regulation of permitted discharges. Organochlorine pesticides have been withdrawn from use and an effort to destroy all existing PCBs by high temperature incineration is in progress.

Attention has been focused particularly on the bioaccumulating metals mercury and cadmium which may enter the human food chain and/or reach top predators such as seals. Concentrations in fish are generally low: in the range 0.03–0.22 mg kg^{-1} for mercury and 0.02–0.66 mg kg^{-1} for cadmium. The higher values are recorded in fish from the southern Bight and the German Bight, including the Thames, Humber, and Ems estuaries. Concentrations of cadmium in the livers of seals from the east coast of England are 0.04–1.0 mg kg^{-1}, and in the kidneys of Scottish seals 0.04–4.0 mg kg^{-1} wet weight, both figures being well below the concentration of 13–15 mg kg^{-1} that is considered hazardous.

Contamination by other metals can be detected in most parts of the North Sea, but concentrations are generally not sufficient to cause concern.

Although DDT has long been withdrawn from use and a downward trend in contamination levels can be detected (Fig. 9.3), high levels still occur at several places and there are clear concentration gradients off the estuaries of the Thames, west Scheldt, Weser, and Elbe. There is thus a continuing input, presumably derived from runoff from agricultural land. PCB contamination of mussels and fish has similarly shown a downward trend, although high concentrations can still be found in fish from the western Scheldt estuary, the Thames estuary, and those parts of the Waddensee influenced by the rivers Ems, Weser, and Elbe.

Fisheries

Since the resumption of fishing after the Second World War, the North Sea has been subject to great fishing pressure. The fishing is monitored and regulated, and catch quotas are agreed each year, depending on the strength of recruitment to the fish stocks, but these quotas are often exceeded, and the intensity of fishing pressure has been responsible for a serious decline of some fish stocks. In the 1960s, mackerel (*Scomber scombrus*) became severely depleted and the spawning stock has remained low ever since. In the late 1970s, the North Sea herring (*Clupea harengus*) fishery collapsed. In the early 1990s, haddock (*Melanogrammus aeglefinus*) and cod (*Gadus morhua*) stocks were at their lowest for thirty years, giving concern about the future of these fisheries also.

The fishery is wasteful because of the large number of fish that are discarded because they are of unwanted species or are undersized. Trawlers in the west and north-west North Sea discard 52 per cent of their catch; in the eastern North Sea seine fishing discards 55 per cent; in the southern North Sea beam trawlers discard 56 per cent of the catch.

An important development resulting in increasing pressure on fish stocks has been the introduction of 'industrial' fishing of small fish for the production of oil and fish meal. Before the 1970s, herring and mackerel formed the basis of this industry, but depletion of these fish stocks resulted in Norway pout (*Trisopterus esmarkii*) and sprats (*Sprattus sprattus*) being used in their place (Norway pout had not formerly been fished). With the decline of these stocks, they were replaced by sand eels (*Ammodytes*) after 1985 and now represent two-thirds of the catch for fish meal.

Although the strength of recruitment varies from year to year, the practices of the fishing industry have a dominating influence on the size of exploitable fish stocks in the North Sea. Against this background, it is difficult to detect any effects that pollution may have on a fishery. An exception is the incidence of fish diseases in some areas that remain contaminated with conservative pollutants. Epidermal papillomas are common on dab (*Limanda limanda*) and other flatfish from former dumping grounds for titanium dioxide waste

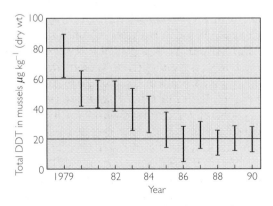

Fig. 9.3 Total DDT concentration in mussels near Dunkirk, France.

in the German Bight and off the Dutch coast, and at the outflows of the River Rhine/Meuse along the Dutch coast, off the Humber estuary, and on the Dogger Bank.

Oil and sea-birds

The North Sea has a large resident bird population and the coasts provide wintering grounds for very large numbers of shore and sea-birds. The dense flocks of birds are vulnerable to oil pollution and there are numerous casualties each winter.

The Straits of Dover are too shallow to allow the passage of a fully laden large oil tanker, so the North Sea has been spared catastrophic oil spills, although it has had its share of tanker accidents. The worst case was in 1955 when the tanker *Gerd Maersk* grounded near the mouth of the River Elbe and spilled 9000 t of crude oil on to mud-flats occupied by roosting waders; estimates of the number of birds killed ranged from 50 000 to 500 000. In February 1969, a few hundred tonnes of fuel oil drifted among the Dutch islands and through the Waddensee, killing 35 000–41 000 birds, including a large proportion of the Dutch breeding population of eiders (*Somateria mollissima*). The largest oil spill was in the 1977 blow-out of a well in the Ekofisk field in the Norwegian sector of the North Sea, in which 20 000–30 000 t of crude oil were discharged, but the oil dissipated without any detectable effect.

Major oil spills such as these are probably of less significance than the chronic small-scale oil pollution from general shipping. The North Sea has been designated a 'Special Area' and it is illegal for ships to discharge oily ballast and bilge water, but these practices continue and are responsible for the large number of small oil slicks that are reported. Quite small quantities of oil may cause heavy losses of sea-birds: 12 000 casualties on the east coast of Scotland in 1971 and 30 000 in the Skagerrak in 1981 were both the result of small oil slicks from unknown sources. Regular counts of dead oiled birds found on North Sea coasts over the last 20–30 years have shown no sign of a reduction in the number of casualties. The greatest number of

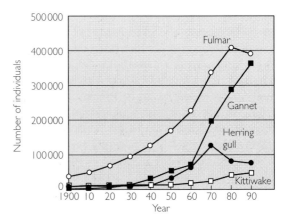

Fig. 9.4 Changes in the populations of some sea-birds in the North Sea.

casualties is in winter, and the prevailing westerly winds drift both the floating oil and birds towards the continental coast, where 70–80 per cent of dead birds on the beaches have been oiled (see Fig. 4.11, p. 54).

Despite these great and continuing losses, earlier fears that they were causing a reduction of the breeding populations of a number of sea ducks (scoter (*Melanitta nigra*), long-tailed ducks, (*Clangula hyemalis*), and so on) and auks (Alcidae), now appear to be unfounded. In fact, many sea-bird populations around the northern North Sea have undergone a considerable increase in recent years (Fig. 9.4), to the extent that some colonies are now limited by the food supply.

More insidious threats to the wintering bird flocks are the reclamation of coastal wetlands and shallows, destroying irreplaceable feeding grounds, and the depletion of the fish stocks on which the birds depend for food. It is not possible to say whether these activities have had any impact on bird populations so far, but the destruction of the food resources cannot be expected to continue without some adverse consequences.

International action

Atmospheric and river-borne inputs account for a considerable fraction of the wastes entering the North Sea, and these are derived from much of western and central Europe, not simply the population of the coastal zone. The River Elbe carries wastes from the Czech Republic and Germany into the North Sea, the Rhine receives inputs from Switzerland, France, Germany, Luxemburg, and the Netherlands; the Scheldt arises in Belgium but flows to the North Sea through the Netherlands. While it is within the power of individual states to improve the quality of some of their waters, in many cases, particularly in areas influenced by the outflow from continental rivers, only international action can achieve any improvement.

In 1974, under an international convention, the Oslo Commission was established to regulate discharges and dumping from ships into the North Sea. The comparable Paris Commission was set up a little later to regulate direct discharges from land. In addition, since most North Sea states are members of the European Union, the European Commission has some power to regulate waste inputs and has acted, for example, by stipulating maximum permissible contamination of shellfish for human consumption, and bacterial contamination of bathing beaches.

The International Council for the Exploration of the Sea (ICES) includes representatives of all the countries fishing the North Sea and for almost a hundred years has been responsible for evaluating fish stocks and recommending annual fishing limits. For the last 25 years it has also had an important role in arranging monitoring programmes and collecting scientific information about pollution levels and impacts.

Since 1984, environment ministers of the North Sea states have met at intervals to agree the extent and source of the more important wastes reaching the sea, and for phased reductions of inputs, particularly those affecting the most vulnerable areas in the eastern North Sea.

THE BALTIC SEA

The Baltic Sea is the largest body of brackish water in the world, with an area of 370 000 km^2. It consists of several basins of various depths, the greatest being 495 m, separated from each other by relatively shallow sills (Fig. 9.5). Communication with the North Sea is through the narrow Öresund between Denmark and Sweden, which is only 7–8 m deep at its shallowest point, and the Belt Sea between the Danish islands of Fyn and Zealand, where the minimum depth is 17–18 m.

The annual input of freshwater from the land varies between 440 and 470 km^3, depending on climatic factors, and the input of salt water from the North Sea is about the same. These inputs are balanced by an equivalent outflow into the Kattegat. Surface salinity is below 4 per mille in the inner parts of the Gulfs of Bothnia, Finland, and Riga, 6–8 per mille in the main part of the Baltic, and about 9 per mille around the southern coast of Sweden. During periods of reduced input of North Sea water, the salinity of the Baltic Sea gradually falls and there is a corresponding southward shift in the distribution of salt water and freshwater species.

The brackish Baltic water floats on denser, more saline, deeper water and, while wind and thermal convection currents produce some vertical mixing in autumn and winter, there is a permanent halocline at a depth of about 40 m in the south-western basins, and at 60 m in the central basins, separating the surface layer from the deeper water (Fig. 9.6).

Baltic water flows out to the North Sea over an inflowing, deeper current of high salinity North Sea water. Under appropriate conditions of wind direction and speed, the flow over the sills in the Öresund and Belt Sea may be entirely of Baltic water flowing seaward, or of North Sea water flowing inward.

The latter condition is important, because it is only during major inflows of North Sea water that the salinity of the Baltic Sea is restored and water in the deeper parts of the basins is replaced. At such times a second halocline develops at a depth of 110–130 m. Once the deep basins have been recharged with salt water, the density of the new bottom water must be substantially reduced by mixing before a new influx can replace it. This may

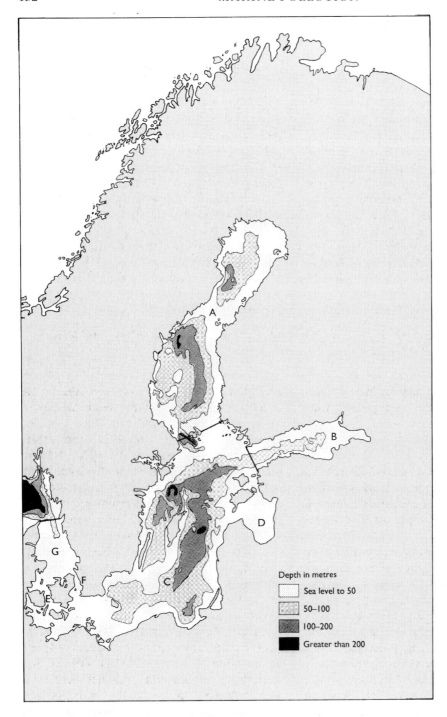

Fig. 9.5 Deep basins of the Baltic contours at 50 m, 100 m, and 200 m. A, Gulf of Bothnia; B, Gulf of Finland; C, Baltic proper; D, Gulf of Riga; E, Belt Sea; F, Öresund; G, Kattegat.

Depth in metres

Sea level to 50

50–100

100–200

Greater than 200

take five years or more and during this period the bottom water is essentially stagnant.

During such periods of stagnation, the deep basins become depleted of oxygen through the bacterial degradation of organic material present in the sediment and drifting down from the surface layer. Extensive areas with anoxic conditions and the development of hydrogen sulphide therefore occur at irregular intervals (Fig. 9.7). At such times, the benthic

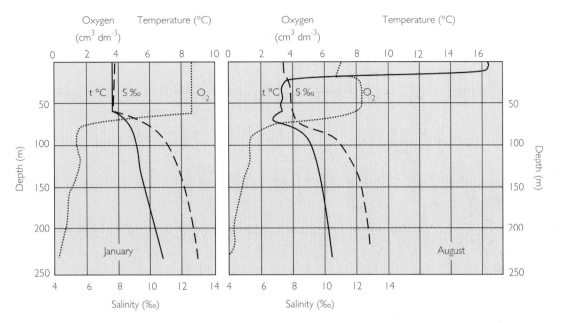

Fig. 9.6 Distribution of temperature (*t*) (*solid line*), salinity (*S*) (*dashed line*), and oxygen (*O*) (*dotted line*) with depth in the Gotland Deep in January and August 1978. (*Elsevier Science*)

fauna disappears until a new influx of North Sea water replenishes the oxygen and allows recolonization.

Eutrophication

In the 1960s when severe and increasing deoxygenation of bottom waters was first detected, there were fears that the Baltic had become catastrophically polluted and was dying.

The Baltic is naturally oligotrophic—it has a small natural organic input and low production—but receives a large quantity of organic wastes and, in a number of areas, coastal waters show signs of eutrophication. Domestic sewage is the most significant organic input in areas such as the south-west Baltic and the Gulf of Finland, particularly the Neva estuary, where there are major centres of population. Intensive fish farming is responsible for a large organic input in the Finnish archipelago. In the Gulf of Bothnia the chief input of organic matter is from the paper and wood pulp industries. There are large inputs of nitrogen and phosphorus, in both organic and inorganic forms, from

agriculture in runoff to rivers; on the south coast of the Gulf of Finland inputs also arise from several fertilizer factories.

The installation of industrial and urban waste treatment plants in the 1970s did not prevent further increases in the inputs of nutrients. In inshore waters primary production by phytoplankton has greatly increased: it trebled between 1970 and 1986 in the outer Gulf of Finland and doubled in Danish coastal waters during the 1970s and 1980s. As a result of this phytoplankton increase, there was a great increase in zooplankton biomass in the Gulf of Riga between 1945 and 1985, and a similar effect has been detected throughout the Baltic. In coastal waters, this increase has been accompanied by a change of species composition of the zooplankton. The relative importance of predatory fish has decreased and, although the total fish biomass has increased, catches in coastal waters, although high, are of less desirable species and so of reduced value.

Less is known about the effect of this enrichment in offshore waters, although heavy phytoplankton blooms particularly of blue–

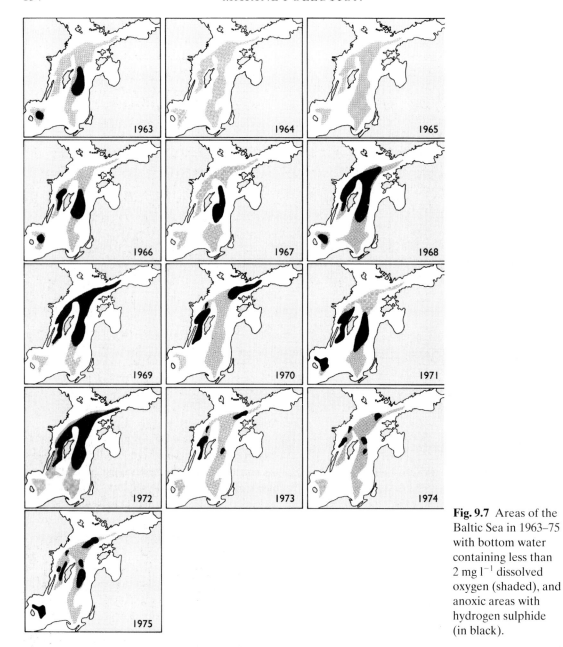

Fig. 9.7 Areas of the Baltic Sea in 1963–75 with bottom water containing less than 2 mg l^{-1} dissolved oxygen (shaded), and anoxic areas with hydrogen sulphide (in black).

green algae, have increased. There are fears that the increased input of detritus to deep water that results from the blooms may accelerate the development of anoxic conditions. To a degree, however, enrichment of the waters is beneficial to fisheries.

The overall yield of fish from the Baltic is fairly low, 2.2 g m^{-2}, compared with 5.2 g m^{-2} in the North Sea, and Baltic fish grow more

slowly. The major incursions of saline North Sea water allow the greater penetration of herring (*Clupea harengus*), sprat (*Sprattus sprattus*), and cod (*Gadus morhua*) into the Baltic Sea, although they do not live in the gulfs where the salinity is below 5–6 per mille. The marine species make the most significant contribution to fish biomass in the Baltic and account for 90 per cent of the landings. The fisheries for migratory eel (*Anguilla anguilla*) and salmon (*Salmo salar*) are economically important, although the total weights landed are small. Commercial freshwater species, perch (*Perca fluviatilis*), pike (*Esox lucius*), roach (*Rutilus rutilus*), and white fish (*Coregonus*), are caught in waters of lowest salinity but are of secondary importance. There is no evidence that eutrophication has damaged these fisheries, but sprat and herring stocks appear to be over-exploited and catches of these species may decline for that reason.

Conservative pollutants

In the 1960s, along with fears about the effects of eutrophication, there was great concern at high concentrations of mercury in fish from some of the coastal and inland waters of Denmark, Finland, and Sweden. The mercury was derived mainly from pesticides, particularly from fungicides and slimicides used in the paper and wood pulp industry in Finland and Sweden, especially in the Gulf of Bothnia. In Sweden, commercial fishing was banned in areas where the fish contained more than 1 mg kg^{-1} of mercury.

In view of these dangerous levels of contamination, the use of mercurial pesticides was phased out, but mercury levels still remain high in sediments in the Gulf of Bothnia and the Stockholm archipelago as a result of former inputs. Mercury concentrations in fish from the Gulf of Bothnia are higher than in those from the Baltic proper, but high levels of mercury are still recorded in fish from some areas in Swedish and Finnish coastal waters, in parts of the Gulf of Finland, and Estonian archipelago, and in the Öresund. Monitoring of mercury contamination of perch from various parts of the Stockholm archipelago failed to show any serious reduction

There has also been a strong effort to reduce direct inputs of other metals. Now, more than 50 per cent of the input of most metals from human activities is by fallout from the atmosphere, and for copper, cadmium, and mercury possibly 80 per cent of the input is atmospheric. Except for mercury, concentrations of metals in sediments have gener-

ally declined since 1980, although high levels of contamination can still be found near centres of population and industry in the Kattegat, Öresund and the Arkona basin, in the Gdansk Bight, and in the Gulf of Bothnia.

During the 1980s, the bottom fauna showed a corresponding decline in contamination by metals, and there has been a recovery of benthic communities in most parts of the Baltic. However, restrictions on fishing still exist in some Swedish inshore waters, and for cod and flounder (*Platichtys flesus*) in certain designated areas near Copenhagen.

The use of DDT, HCH, and PCBs has been discontinued in the Baltic states, but these organochlorines can still be detected in sediments and at various trophic levels throughout the Baltic Sea. Fish-eating birds appear to biomagnify dioxins and PCBs and have much higher levels than in fish (Fig. 9.8). The concentration of dioxins in white-tailed eagle (*Haliaetus albicilla*) eggs from the Baltic proper is 27 times as great as that in pike, the major food source of these birds. Guillemot (*Uria aalge*) eggs contain 13 times as much dioxin as found in herring.

Following the ban on DDT, PCBs, 2,4,5,-T, and pentachlorophenols in the 1970s, there has been a 90 per cent reduction of DDT concentrations in fish and birds; PCB levels have declined more slowly, but are now at 50 per

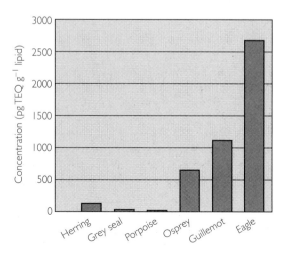

Fig. 9.8 Concentration of dioxins and furans (as TCDD equivalents) in Baltic organisms.

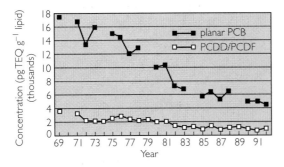

Fig. 9.9 Concentration of dioxins and furans (as TCDD equivalents) and PCBs in guillemots on the coast of Gotland, 1969–92.

cent of their former levels. The steady decline in PCB and dioxin contamination of guillemot eggs has been monitored between 1969 and 1992 (Fig. 9.9) and this has paralleled the decline of the concentration of these contaminants in Baltic herring.

International action

The Helsinki Convention, which came into force in 1980, is concerned with the protection of the Baltic Sea. All eight states with Baltic coastlines are active members of the Helsinki Commission which puts the Convention into effect. Like the Oslo and Paris Commissions for the North Sea, the Helsinki Commission regulates discharges and dumping from ships and direct discharges from land. It also has wider responsibilities and has been concerned with such matters as atmospheric inputs, emergency action against oil spills, the use of antifouling paints containing tributyltin, co-ordinating monitoring programmes, and seal conservation. It frequently arranges meetings of experts to evaluate the scientific evidence on which its recommendations are based. The Helsinki Commission has proved to be very effective: through its efforts the Baltic is, with the North Sea, one of the most thoroughly studied seas in the world.

THE MEDITERRANEAN SEA

The Mediterranean Sea is a deep, virtually tideless sea with a surface area of 2 965 000

km². Much of it is more than 200 m deep, with a number of deep basins below 3000 m. The eastern and western Mediterranean, separated by the relatively shallow straits between Sicily and Tunisia, show differences in resident fauna and flora, indicating a degree of isolation between the two regions. The Aegean, and particularly the Adriatic, are semi-enclosed extensions from the main body of the Mediterranean.

Evaporation is about three times greater than the input from precipitation; the deficit is made good largely by an inflow of Atlantic water at Gibraltar. There is negligible input to the eastern Mediterranean and, as a result, the salinity in the west is near oceanic levels, approaching 37 per mille, but near the coast of Asia Minor it is 39 per mille. This dense saline water sinks to a depth of 100–200 m, flows westwards, and leaves the Mediterranean as a deep current at Gibraltar. Through most of the year there is good vertical mixing of the top 200 m, and sometimes 600 m, of water.

The main centres of coastal population are on the northern side of the western Mediterranean and around the head of the Adriatic (Fig. 9.10). These are also the most industrialized areas and receive the majority of the 100 million tourists that inflate the summer population of Mediterranean countries. The North African coast, by contrast, is for the most part arid, with little urbanization or industrialization. Pressures on the marine environment therefore vary widely depending on the local and human environment.

Fisheries

The Mediterranean is only moderately productive and the demand for fish in Mediterranean countries substantially exceeds the local supply. The deficit is only partly compensated for by imports from the Atlantic and other fisheries, and the cash value of the Mediterranean fisheries is therefore high, with prices several times greater than average world prices. Bottom-living species, such as mullet (*Mugil*), hake (*Merluccius merluccius*), and so on, are most in demand. The main stocks of these are along the northern coast

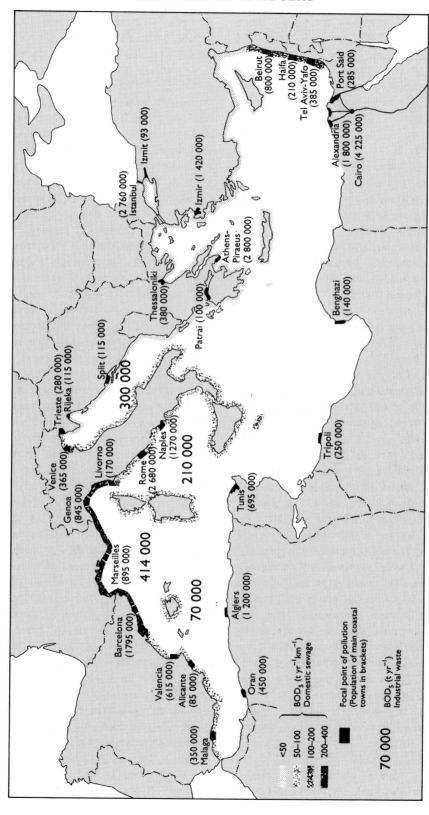

Fig. 9.10 Sewage and industrial waste discharges to the Mediterranean. (*Elsevier Science*)

and these are fully exploited. A number of pelagic fish (for example, anchovy (*Engraulis encrasicolus*), sardines (*Sardina pilchardus*), and mackerel (*Scomber scombrus*) in the north-western Mediterranean) are also intensively fished. Mariculture is practised in some places but is not widespread.

Oil pollution

All parts of the Mediterranean are chronically polluted with tar balls, a high value of 500 l km^{-2} has been recorded south of Italy, and many tourist beaches are irritatingly contaminated with specks of tarry oil. The main sources of oil pollution are deballasting and tank-washing operations of oil tankers and the discharge of oily bilge water by other shipping. Refinery wastes are a secondary, but important source of chronic oil pollution. Offshore oil extraction and tanker accidents have been of minor significance. Some 250 million t year^{-1} of oil are transported through the Mediterranean, about 150 million t year^{-1} of this from North Africa to European ports (Fig. 9.11). The cross-Mediterranean journey time is too short to allow efficient operation of the load-on-top system (see p. 40), and there have been inadequate slop reception facilities in ports. A ban on the discharge of oily wastes in 1976, and stricter enforcement, proved effective and the quantity of floating tar in the eastern Mediterranean declined from 37 000 μg m^{-2} in 1969 to 1175 μg m^{-2} in 1987.

As a result of oil pollution, tainting of a variety of fish and bivalves, rendering them unmarketable, has been reported from the neighbourhoods of oil ports in Spain, France, Italy, and the former Yugoslavia. Spiny lobsters (*Palinurus elephas*) have been killed by oil pollution around Bizerte in Tunisia. The Bay of Izmir and the Sea of Marmora, on the Turkish coast, have suffered adverse effects from oil, and the spawning grounds of bonito (*Sarda sarda*) and mackerel have been damaged. The most serious effects of oil pollution have been noted in the Gulf of Naples, Caligari, and the Venetian lagoon, where fish populations have been reduced. The Bay of Muggia at Trieste, once rich in fish, is now described as being biologically almost a desert because of the impact of petrochemical wastes.

Domestic wastes

A survey made in 1972 found that most coastal communities discharged untreated sewage to the sea and the rest usually gave only minimal treatment. The worst affected areas were the Riviera coast from the River Elbo in Spain, across the French Riviera, to the River Arno in Italy (Fig. 9.10). It was also estimated that the discharge amounted to 336 t km^{-1} year^{-1}. The Israeli coast was found to be similarly affected. A vigorous attack on this problem was mounted, especially in France and Israel and on parts of the Spanish and Italian coast, but it is doubtful if the situation has improved much elsewhere. One difficulty is the need to provide sufficient sewage treatment facilities to cope with the enormous influx of population for a few summer months.

Apart from damage to the tourist industry for aesthetic reasons, discharges of untreated sewage are a threat to public health. In 1973, there was an epidemic of cholera centred around Naples, spread by the consumption of infected shellfish, and lesser gastric disorders have been a common hazard to visitors to Mediterranean coasts for many years.

Eutrophication

The organic load (BOD) from rivers equals that from coastal towns, and most of the phosphorus and nitrogen inputs are from rivers. Phosphorus is the limiting nutrient for phytoplankton in the upper Adriatic and inputs of phosphorus to the area have increased markedly. The upper Adriatic now receives about 30 000 t year^{-1} of phosphorus, of which about 60 per cent is contributed by the River Po. Half the phosphorus is derived from detergents and agricultural fertilizers which came into intensive use after 1945. The progressive eutrophication resulting from this is shown by the increased oxygen concentration in surface waters and oxygen depletion of bottom waters, which first became apparent

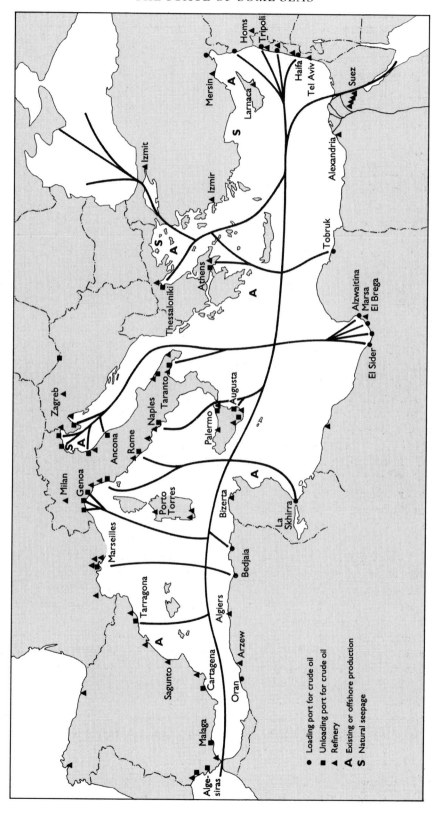

Fig. 9.11 Activities of the oil industry in the Mediterranean. (*Elsevier Science*)

around the mouth of the River Po in the mid-1950s, but now extends to a much wider area in the western and northern parts of the upper Adriatic (see Fig. 3.16, p. 36).

Other areas that suffer eutrophication include the lagoon of Venice, the Gulf of Tunis, Lake Maryut at Alexandria on the Egyptian coast, and, on a smaller scale, numerous marinas where domestic waste from the boats is a significant source of organic matter in waters that are often stagnant.

Conservative pollutants

There is some uncertainty about the relative importance of atmospheric inputs of conservative pollutants. At one time it was thought that 85 per cent of the anthropogenic inputs to the Mediterranean were by rivers and direct discharges. It now appears more likely that atmospheric inputs of lead, zinc, chromium, nickel, and mercury to the western Mediterranean are as great as, or greater than, other inputs, and the atmospheric input of organochlorine pesticides is also thought to be large. However, there is a considerable input of pesticide residues, heavy metals, and PCBs from the rivers Rhone and Po, adding to inputs from the heavily industrialized zones around their mouths. High concentrations of metals and PCBs occur in sediments around outfalls from Marseille, Nice, Naples, Athens, and other major cities.

Areas that have been studied in detail include the Gulf of Fos influenced by the Rhone and its industrial belt, the upper Adriatic, and the Saronikos Gulf which receives the effluents of Athens and Piraeus. All show some environmental degradation resulting from the effluent load they receive. There is most concern about the conditions in the Venetian lagoon and upper Adriatic which, because of their low water exchange and relative isolation from the rest of the Mediterranean, are least able to accept a heavy load of effluents without damaging consequences.

Contamination of organisms

Despite the existence of areas where conservative pollutants accumulate in undesirably high concentrations, neither the water nor the organisms in the Mediterranean appear to be seriously contaminated, and, with the exception of mercury, concentrations are comparable to those in the open Atlantic.

Parts of the Mediterranean receive a natural input of mercury from mercury-bearing ores and volcanic activity, and this is reflected in elevated mercury levels in some marine organisms. Mercury levels in two species of tuna are 2–3 mg kg^{-1}, three times greater than those in Atlantic tuna. Fish from the northern Tyrrhenian Sea, off the coast of Tuscany, also have high concentrations of mercury: larger sized hake, poor-cod (*Trisopterus minutus*), an octopus (*Eledone cirrhosa*), and Norway lobster (*Nephrops norvegicus*) all contain more than 1.0 mg kg^{-1} mercury (up to 3.2 mg kg^{-1} in hake).

There have been instances where seafood from areas receiving waste discharges has been contaminated. In the early 1970s, a few fishing communities on the Italian Adriatic and French coasts were found to be exposed to high levels of mercury resulting from local inputs contaminating seafood organisms in the area. These inputs were subsequently brought under control. Mussels (*Mytilus*) from the coast near Marseille contained high concentrations of copper (95 ppm), and high concentrations of zinc (200 ppm) have been recorded in mussels from a number of areas, but neither constitutes a health risk.

There is evidence that PCBs in the Mediterranean are subject to 'demagnification'. Microplankton strongly accumulate PCBs in the areas with a large local input, but contamination of animals at higher trophic levels is progressively reduced. The faecal pellets contain ten times the concentration of PCBs found in the animals that produce them and when the pellets sink to the seabed in the deeper parts of the sea they are removed from the pelagic food chains. The same phenomenon has been reported for radioactivity.

Conservation

A dominant feature of considerable areas of the Mediterranean, particularly its northern

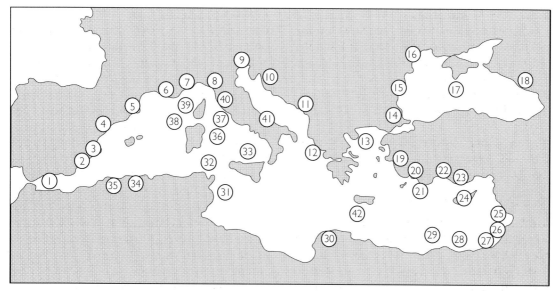

Fig. 9.12 Marine parks and nature reserves in the Mediterranean.

coasts, is the impact of tourism. Tourist resorts have been developed in formerly pristine areas and in summer, sandy beaches, and the shallow sea off them, attract dense crowds. Even rocky coasts suffer continual disturbance from skin-divers, boating, and water sports, and the depredations of spear fishing.

Reconciling mass tourism with conservation of the natural environment and other legitimate uses of the coast and coastal waters is not easy. One approach to this has been by reserving certain critical coastal and sea areas for particular purposes.

Marine reserves of various kinds have been established in Spain, France, Monaco, Italy, the former Yugoslavia, Greece, Israel, Egypt, and Tunisia (Fig. 9.12), and more are planned. Most are concerned with fishery enhancement, but a number are marine nature reserves which provide comprehensive protection whilst allowing for educational, scientific, and tourist activities. In several cases marine reserves have specialist laboratories associated with them and provide facilities for research in the protected areas.

The Greek marine reserves on Zante and in the northern Sporades are important because they safeguard breeding areas of the endangered monk seal (*Monachus monachus*), and

the sandy beaches of the Sporades provide undisturbed areas where the turtle *Caretta caretta*, which is also an endangered species, comes ashore to lay its eggs.

International action

Many of the problems of the Mediterranean are local in character and are susceptible to local solutions. While the financially important fisheries have not been generally affected by pollution and fish catches continue to grow, the Mediterranean receives a very heavy effluent load which, because of restricted water exchange, poses a continuing threat. The United Nations Environment Programme (UNEP) has provided a focal point for international cooperation in cleaning up the Mediterranean, and this culminated in the 1976 Barcelona Convention to put this into action.

Following this, oil pollution has been much reduced and a great deal of information has now been collected identifying inputs and levels of contamination. Progress in other remedial action has been slow for a variety of reasons. There are 21 Mediterranean states (or 26 if Black Sea states are included) and it is difficult to achieve concerted action by so many countries. The problem is compounded by severe political obstacles in some cases, and by the great diversity in the degree of economic development of different coastal states, with corresponding differences in the priority accorded to environmental protection.

THE CARIBBEAN

The greater Caribbean area includes the Caribbean Sea and the Gulf of Mexico. Its total surface area is 4.24×10^6 km^2, and it consists of a number of deep basins separated by major sills. It lacks a shallow continental shelf, and except off the US Gulf coast, in the Yucatan Gulf, and part of the South American shelf, 80 per cent of the water is deeper than 1800 m and half is deeper than 3600 m, with the greatest depth of 7100 m in the Cayman Trench.

The Caribbean Sea and southern half of the Gulf of Mexico are tropical and experience little seasonal change. The surface water temperature is 27°C with a seasonal fluctuation of less than 3°C. The northern part of the Gulf of Mexico, however, has a winter surface sea temperature of 16°C, rising to 28°C in summer. Over much of the area there is a permanent thermocline at a depth of about 100 m and upwelling is not a dominant feature, although there are some local areas where bottom water comes to the surface. A plume of freshwater from the Orinoco river extends in a north-westerly direction and can be identified as far as Puerto Rico. It introduces plant nutrients and is marked by a phytoplankton bloom along its course.

Otherwise, because of the thermocline and lack of upwelling, much of the area is deficient in nutrients. For this reason, significant fisheries, particularly for penaeid prawns, are confined to the shallows of the Mexican and United States coasts. These are associated with the highly productive coastal ecosystems of mangroves, seagrass, and coral reefs. The US commercial fishery in the Gulf of Mexico is worth $400 million per year, and the income from recreational fishing is at least as great.

Seagrass beds (Fig. 9.13), principally of turtle

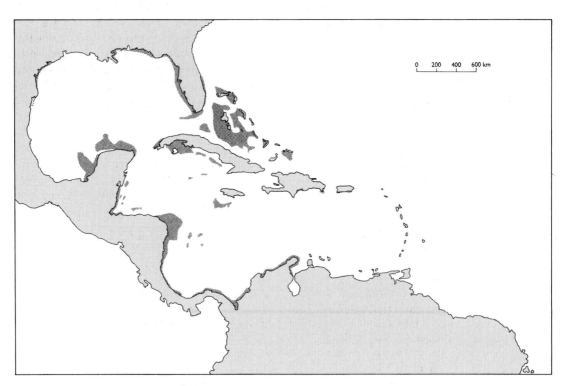

Fig. 9.13 Distribution of seagrass beds in the Caribbean. (*Elsevier Science*)

grass (*Thalassia testudinum*), can contribute 3000 g m^{-2} year^{-1} (dry weight) of primary production, of which two-thirds is due to sea-grass growth and one-third to the associated algae. This is the same as, or even greater than, the production of mangrove swamps. On coasts fringed by mangroves, with seagrass beds beyond them and coral reefs to the sea-ward side, very high levels of production are reached. Fish leave the shelter of the reef to feed in the seagrass beds, and both seagrass and mangroves, besides harbouring a large and varied fauna, provide important nursery grounds for prawns and fish.

Another feature of mangroves and seagrass beds is that their root systems stabilize the seabed and prevent coastal erosion. They also clarify the water by trapping sediments.

In the Caribbean region as a whole, there is relatively little coastal urbanization or in-dustrialization. Apart from the Orinoco plume, which probably carries contaminants at low concentrations to the entire eastern Caribbean, and oil, there is no widespread pollution. In a number of areas, however, there have been severe local effects from a variety of pollutants and other causes (Table 9.1).

Oil pollution

The principal oil-producing areas are Trinidad and Tobago, Venezuela, and the Gulf of Mexico. Nearly one-third of the production is from offshore oilfields and there are more than 2000 fixed offshore platforms in the US sector of the Gulf alone. Blow-outs, overflows, pipeline fractures, and other accidents at the platforms are a major source of oil pollution in the area.

Studies of the benthic fauna in oilfields off the Louisiana coast have failed to reveal any significant differences from other areas, and the platforms themselves, providing the only hard structures in a generally sedimentary area, attract an exotic flora and fauna as well as sheltering a variety of sport fish. Despite the spillages and some contamination of the sediments, offshore activities do not appear to have a serious environmental impact.

One of the largest oil spills in history was that following the blow-out of the Ixtoc 1 field in the Bay of Campeche off the Mexican coast. From June 1979 to March 1980, when the well was finally capped, some 350 000 t of crude oil were released into the sea. Prevail-ing offshore winds prevented fresh oil coming ashore on the Mexican coast, but 80 km of the Texas coastline, some 800 km away, were seriously affected. Entrances to the environ-mentally sensitive Laguna Madre were boomed off to protect the breeding grounds of the endangered brown pelican (*Pelecanus occidentalis*). Ten thousand baby Kemp's

Table 9.1 Incidence of severe local pollution damage in the Greater Caribbean

Location	Pollutant
Cartagena, Colombia	Mercury
Colon, Panama	Oil
Veracruz, Mexico	Thermal effluent
Galveston, Texas	Oil
New Orleans, Louisiana	Industrial waste
Tampa Bay, Florida	Sewage
Kingston, Jamaica	Bauxite waste
Port-au-Prince, Haiti	Sewage
Santo Domingo, Dominican Republic	Urban run-off
San Juan, Puerto Rico	Sewage
Port of Spain, Trinidad and Tobago	Oil
Cumana, Venezuela	Sewage
Havana Bay, Cuba	All the above

Ridley turtle (*Lepidochelys kempii*), another endangered species, were airlifted from the only breeding ground of this species on beaches on the north coast of Mexico to another area of the Gulf of Mexico that was free from the threat of oil. Despite the enormous quantity of oil spilled, the environmental consequences were remarkably small.

There has been some controversy about the vulnerability of reef corals to oil pollution, but there is increasing evidence that at least some species are susceptible. Branching corals, such as *Acropora palmata*, are most sensitive to human disturbance, but some of the massive corals are much less so. The impact of oil pollution is therefore not uniform. Downstream of a large refinery at Aruba on an otherwise uniform coast, fringing reefs have a reduced species diversity and less cover of scleractinian corals, particularly *Acropora palmata* and *Montastrea annularis*. Massive corals are less affected. Studies made after a large oil spill from a ruptured storage tank at a refinery near the Panama Canal in 1986 also showed that numbers of corals, total coral cover, and species diversity were significantly reduced with increasing amount of oiling. The massive coral, *Siderastrea sideria*, showed no effect of oil on growth in the two years following the spill, unlike *Porites asteroides*, *Monastrea annularis*, and *Diplora strigosa*, which all showed sharp reductions in growth.

Sedimentation

Turbid water reduces the productivity of seagrass beds by restricting light penetration, and heavy sedimentation is lethal to corals. The Mississippi and a number of lesser rivers flow into the Gulf of Mexico, and most of the water from the Orinoco travels north and enters the southern Caribbean. These rivers carry a heavy sediment load, particularly from areas where bad farming practices result in soil erosion. The damaging effect of these inputs is limited to the areas, sometimes large, around the mouths of the rivers.

On several islands, clearing of mangroves and deforestation of hillsides has led to soil erosion and increased runoff of sediments, particularly following heavy rain. The increased sedimentation is said to have smothered corals at two places and is certainly detrimental to coastal fisheries.

More widespread damage is caused by coastal developments involving dredge and fill, or the dredging of shipping channels. Dredging operations cause increased turbidity of the water, and marine dumping of the spoil spreads the damage over a wider area.

The common practice of extracting beach sediment for construction purposes, as well as coastal developments, largely for the tourist trade, leads to the destruction of mangrove forests and seagrass beds, and has caused severe coastal erosion in a number of places. This, in turn, leads to increased turbidity of the water and increased sedimentation elsewhere.

Heated effluents

The sea temperature is close to the upper thermal limit for many organisms and they are therefore vulnerable to thermal pollution. This is not a widespread hazard because of the low level of industrialization of coasts in the Caribbean, but cooling water discharges from two power stations in Puerto Rico have caused extensive damage to seagrass beds in Guayanilla Bay. Plants are killed and production is reduced in areas where the sea temperature is 5 °C above the ambient temperature of 30 °C. The benthic fauna in shallow waters is completely eliminated at this, or higher, temperatures.

Nutrients

Caribbean waters are naturally low in nutrients so that even slight increases from sewage pollution may be expected to lead to serious environmental impacts, particularly on coral reefs. While such damage has been recorded in a number of areas, it is not always easy to separate the effects of nutrient inputs from the effects of over-fishing and other influences. In one detailed study in the Florida Keys, it has been found that, although many inshore areas near coastal developments do have elevated nutrient and chlorophyl *a* levels, most of the nutrients entering the coastal waters from shore are taken up by algal and seagrass communities before they reach the reef areas.

Coral reefs

Coral reefs are an important feature of the marine environment in the Caribbean, but in many areas they have been in decline for several decades. A reduction from 70 per cent coral cover in 1977 to less than 5 per cent in 1993 has been recorded in Discovery Bay Jamaica, and in the same period, macroalgal cover increased from 1–3 per cent before 1983 to over 90 per cent in 1993 (Fig. 9.14).

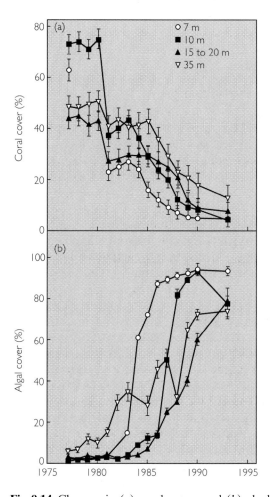

Fig. 9.14 Changes in (a) coral cover, and (b) algal cover between 1975 and 1995 near Discovery Bay, Jamaica. (*American Association for the Advancement of Science*)

The causes of such dramatic changes in ecosystem structure are undoubtedly complex and may be natural or man-made (Fig. 9.15). Hurricanes can be responsible for severe damage, when the force of heavy swells is sometimes sufficient to reduce branching corals to rubble, overturn head corals, and may even raze entire reef tracts to the base substratum.

Although the underlying causes are unknown, disease may also be responsible for major changes. White band disease and black band disease that affect corals in many areas make them more susceptible to attack from carbonate-boring organisms, leading to their collapse. The mass die-off of echinoids *Diadema antillarum* in Discovery Bay was probably a result of disease and is likely to have been responsible for the excessive growth of macroalgae in the absence of these dominant herbivores.

Man-made damage by anchors and ships running aground may seem minor compared to hurricane damage, but is still locally significant. A 20 per cent destruction of *Acropora cervicennis* at the Dry Tortugas, Florida, has been caused by shrimp boats anchoring. Other damaging influences are runoff, the use of chemicals to capture lobsters, increased nitrates and phosphates, over-fishing, coral collecting, dynamiting, and people breaking coral while walking on the reef flats, diving, or swimming. Chemicals in runoff from sugar and banana plantations pollute reef waters off Guadeloupe, and siltation around a bauxite loading pier causes reef damage in Jamaica.

Tourism

The greater Caribbean receives over 100 million tourists annually and tourism represents a major industry, particularly in the eastern islands. The attractions are sun, white sandy beaches, and warm clear waters. Hotel development is almost entirely on the coasts and often in isolated places. There is a general lack of awareness about the environmental impact of such developments, and while some hotels install their own sewage treatment plants, many do not, or have inadequate treat-

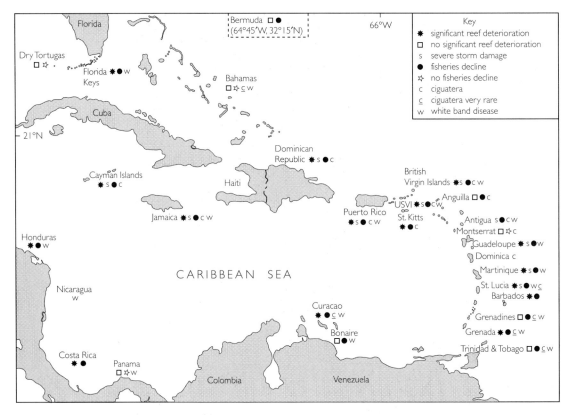

Fig. 9.15 Coastal conditions in Caribbean states.

ment facilities. If, as is commonly the case, the coastal developments have resulted in the loss of seagrass beds and mangroves leading to increased turbidity of the water, the combination with sewage pollution is detrimental to the interests of the tourist industry, even though the effects are quite local.

International action

Under the Regional Seas Programme of the United Nations Environment Programme (UNEP), an 'Action Plan' for the greater Caribbean was adopted by twenty-two states of the region in 1981. UNEP provides the co-ordinating secretariat for the plan. Since then, a survey has been made of the special needs and problems of the Caribbean, and in 1983 a Convention was adopted by the major coastal states which requires the signatories to: prevent, reduce, and control pollution; safeguard sensitive environments; and collaborate in sci-entific research and monitoring. Particular emphasis is placed on oil pollution and a convention for the protection of endangered species is under discussion.

THE CASPIAN SEA

The Caspian Sea (Fig. 9.16) is by far the world's largest land-locked body of water. The northern part is very shallow, only 10–12 m deep, but the central Caspian has depths to 700 m and the southern part, separated from it by shoals in the region of the Apsheron peninsula, has depths to 1000 m.

The surface sea temperature is 24–27 °C in summer, but in winter it falls to 9 °C in the south and 0 °C in the north where extensive sea ice is formed. There is little variation of salinity with depth, but if differs widely in different parts of the Caspian. In the northern shallows it is 5–10 per mille, falling to 2 per

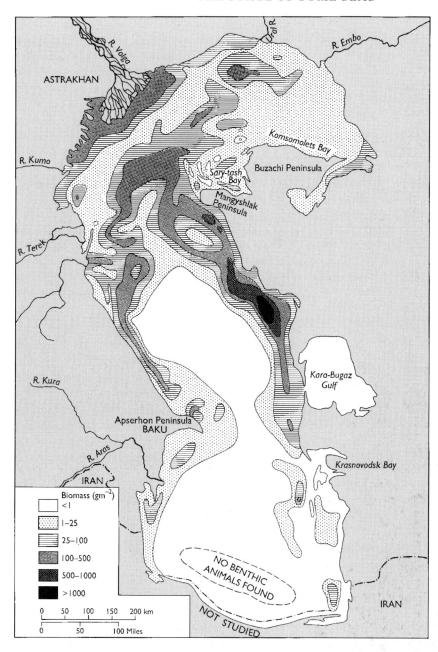

Fig. 9.16 Distribution of benthic biomass in the Caspian Sea.

mille near the Volga delta, and in the central and southern parts it is 12–13 per mille. The eastern side of the sea receives little precipitation, has negligible river input, and the rate of evaporation is high. As a result, bays and semi-enclosed areas of sea in these areas have very high salinities, reaching 200 per mille in the Kara–Bugaz Gulf which is almost com-

pletely separated from the Caspian Sea proper. The composition of the salt differs from that in the open oceans, with more magnesium, calcium, and sulphate, and less sodium and chloride.

A peculiar feature of the Caspian Sea is that it experiences significant changes in sea level (Fig. 9.17(b)). Sea level is referred to a

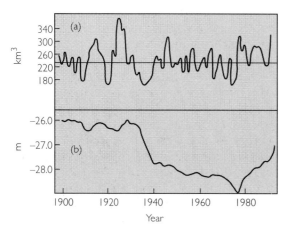

Fig. 9.17 (*a*) Annual input of water to the Caspian Sea from the River Volga; (*b*) mean sea level of the Caspian Sea. (*Elsevier Science*)

datum based on the long-term, average sea level of the Baltic, and historical records since 1500 show that at various times it has varied between −22 m and −29 m. In this century, between 1900 and 1929, the sea level showed minor fluctuations around −26.2 m, but in the 1930s drought conditions resulted in a dramatic 1.8 m fall between 1930 and 1941, followed by a slower decline to −29.0 m in 1977, the lowest level recorded. In the 1940s and 1950s dams for power generation and irrigation schemes were constructed on all the rivers entering the Caspian, except for the Ural. The reduction of river inputs, most particularly that of the Volga (Fig. 9.17(a)), that this caused was responsible for about one-third of the total sea-level fall of about 3 m since the 1930s.

This downward trend was abruptly reversed and after 1978 there was a rapid rise in sea level to −27.0 m in 1992, a 2 m rise in 14 years. This rise in sea level has had the greatest impact on the shallow northern sector, where it has resulted in an extension of the boundaries of the Caspian Sea, particularly to the north-east where the flood plains in Kazakhstan have been inundated.

Fauna of the Caspian Sea

Despite the existence of a similar range of habitats and broadly similar conditions, the fauna of the Caspian Sea is impoverished compared with that of the Black Sea, with only 40 per cent of the number of species. Nearly half the species are endemic and over two-thirds of the species are found only in the Caspian Sea and Aral–Black Sea basin. Since the 1920s, and particularly since the opening of the Don–Volga canal in 1954, more than thirty species have appeared in the Caspian as a result of accidental or deliberate introductions.

Many benthic ecological niches are not fully exploited in the Caspian Sea and in some cases are not used at all. Following the discovery in the 1930s, that large areas of the inshore, soft substratum were virtually uninhabited, the polychaete *Nereis diversicolor* and the bivalve mollusc *Abra ovata* were successfully introduced in large numbers in 1939–40 and 1947 to improve the food supply for sturgeons which are very important commercially. These introductions have become well established and have not affected the native fauna because they have occupied the vacant niche of detritus feeders living in soft bottoms.

There are 126 species and subspecies of fish in the Caspian, half of them endemic. The clupeids and some gobies are fully marine, but most species are associated with the rivers. These include many species that feed and breed in and near the river deltas, and anadromous species, such as the sturgeons, salmon, and lampreys (*Petromyzon*), which formerly moved hundreds of kilometres upstream to spawn but are now more restricted by the dams.

The coasts and open water of the Caspian support several hundred species of shore birds, waterfowl, and sea-birds. About half the ducks, geese, and waterfowl of the former Soviet Union winter in and around the Caspian, and the area is an important staging post for migrants. The deltas of the Volga (10 000 km^2) and the Ural (600 km^2), coastal lakes, and lagoons, as well as the coastal shallows, all provide vital habitats. Rare and endangered species include flamingo (*Phoenicopterus roseus*), red-breasted goose (*Branta ruficollis*), mottled teal (*Anas angustirostris*), and the redbill (*Platalea porphyrio*).

The Caspian seal (*Pusa capsicus*) is the only aquatic mammal. It mates and pups on the sea ice of the northern Caspian in winter, where the chief mortality from predators takes place. It is distributed throughout the Caspian Sea for the rest of the year. It was intensely hunted, but a rapid decline in numbers in the 1950 and 1960s led to the introduction of new hunting regulations and hunting quotas. Seal numbers have now stabilized at 400 000–500 000 individuals.

Ecosystem fluctuations

Changing habitat conditions associated with changes in sea level have had a major impact on the benthic fauna. The fall in sea level and corresponding increase in salinity during the 1930s led to a crash in the average biomass of the benthic fauna of the northern Caspian from 36.84 g m^{-2} to 4.74 g m^{-2}. However, the increase in sea level since 1977 has restored the situation and the distribution of species such as *Nereis diversicolor* has extended into the new areas that are now submerged.

Quite apart from the changes induced by long-term sea-level fluctuations, the benthic fauna is also subject to wide fluctuations from year to year and season to season. In part this is owing to the continuing instability of populations of immigrants and the native species they compete with, and to grazing by sturgeon. The biomass of many organisms (including *Nereis*, *Abra*, and *Corophium*) in the northern and middle Caspian is correlated with the flow volume of the Volga in the previous year, while the distribution of *Nereis* in the northern Caspian is influenced by this factor as well as by changes in sea level.

The feeding habits of two species of sturgeon in the Caspian have an important impact on the benthos. The Russian sturgeon (*Acipenser guldenstadti*) and the stellate sturgeon (*A. stellatus*) feed by sucking up bottom sediments and all the organisms in it are filtered out by the gills. Intensive grazing by these sturgeons can seriously deplete the bottom fauna because small species and juveniles are removed as well as the larger species that are the sturgeons' preferred food. Other bottom-feeding fish include bream, Caspian roach, and goby, but, compared with sturgeon, their impact is insignificant.

In summer, grazing by sturgeon has little impact because the fish are widely dispersed over the extensive shallows of the northern Caspian and in the Volga and Ural rivers. In September and October the sturgeons migrate to the central and southern Caspian (Fig. 9.18). The fish feed down to a depth of 50 m but, because of the narrowing of the shelf in these part of the Caspian, there is intense grazing pressure and the benthic biomass declines by 30–50 per cent, or by 90–95 per cent of the preferred food in some areas. Recovery of these denuded areas does not follow a pre-

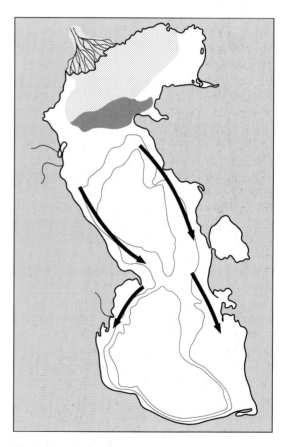

Fig. 9.18 Distribution of migrant sturgeon in April–November (shaded), denser shading indicates the distribution in October–November. Arrows mark the autumn migration route. (*Elsevier Science*)

dictable course and the bottom communities undergo constant readjustments.

Fisheries

The once productive fisheries of the Caspian Sea have been adversely affected by a combination of factors. These include the fall of sea level and associated increase in salinity, the damming of the rivers which has reduced river flow and restricted the spawning areas of anadromous species, and over-fishing. Although the annual catch has remained steady at nearly 400 000 t year^{-1}, the yield of valuable species has decreased fivefold, but the catch of the small clupeid kilka (*Clupionella delicatula* and *C. engrauliformis*), used for fishmeal, has increased dramatically.

The most important commercial fishery is for sturgeon, three species of which are currently taken: the beluga (*Huso huso*), the Russian sturgeon, and the stellate sturgeon. The reduction of their spawning grounds by the construction of dams and severe over-fishing has brought the fishery to a low level several times. The imposition of fishing quotas, increasing the minimum size of the catch, and prohibiting fishing for sturgeon in the open sea has had only a temporary success and in recent years illegal fishing for sturgeon at sea has continued on a large scale. The annual catch dropped from about 27 000 t year^{-1} in the mid-1970s to 10 000 t in 1978 and 2000 t in 1992. Beluga, which represented 40 per cent of the catch in the early twentieth century, now forms less than 10 per cent of the catch; Russian and stellate sturgeon make equal contributions to the rest of the harvest.

In the 1970s a number of fish farms were established to produce sturgeon fingerlings and so enhance the fishery. Thirteen fish farms have the capacity to produce 100 million fry per year. In the mid-1990s, however, the Volga fish farms had difficulty finding enough mature adults to maintain the breeding programme. This was because many of the natural spawning grounds were dry during the period of extremely low river flow around 1977, and few fry survived, resulting in a shortage of mature fish in the 1990s.

Other anadromous species have suffered as a result of the dam construction. Catches of salmon (two species) and shad (nine species) have both been drastically affected. The inconnu salmon (*Stenodus leucichthys*) had its 3000 km ascent of the River Volga for spawning reduced to 450 km. Salmon catches fell from 1500–1900 t in 1938–9 to 10–20 t in 1980. The annual catch of shad fell from 130 000–160 000 t in 1913–6 to 10 t by the early 1960s.

Pollution

The principal source of pollution in the northern Caspian is the River Volga. About 120 km^3 year^{-1} of domestic and industrial waste water is discharged from fifteen towns situated on the river and its tributary the Kama, and a further 8 km^3 year^{-1} is discharged by coastal towns. Of this, 2.5 km^3 is untreated and 7.5 km^3 receives only primary treatment. Metal concentrations in Volga water have increased dramatically, with copper at 7.0 µg l^{-1}, zinc at 22.5 µg l^{-1}, lead at 1.3 µg l^{-1}, and cadmium at 0.5 µg l^{-1} in the late 1980s. Organochlorine pesticide concentrations of 1.8–1.9 µg l^{-1} are also high. Metals and pesticides have contaminated bottom sediments off the Volga delta, but no impact of this has so far been detected.

Although there is some oil contamination in the Volga delta area derived from industry at Astrakhan, oil pollution is not generally a serious problem in the northern Caspian. Sediments on the eastern coast, however, are highly contaminated with oil. The Tengiz oil-field in Kazakhstan is close to the coast and storm surges flooding the coastal oil and gas fields carry oil back into the sea. Offshore oil and gas deposits have been discovered in the northern Caspian Sea and are likely to be exploited.

There are many natural oil seeps in the southern Caspian, and the Azerbaijan coast is the site of a major oil industry with numerous offshore oil wells, together with refineries and petrochemical plants on the coast. This development has been accompanied by serious oil pollution. In the mid-1960s it was estimated that 1 million t year^{-1} of oil and petroleum products were lost to sea from accidental spillages, seepages, pipeline failure, shipping, industrial effluents, and refinery waste water. In the 1970s and 1980s, oil sheens were reported off most of the western coast of the southern Caspian and as far as the eastern coast near the Apsheron peninsula. Tar balls are common in bottom sediments near operating wells and, in several areas, sediments are severely contaminated with petroleum hydrocarbons: 27 mg kg^{-1} around Sumgait, 46 mg kg^{-1} in the Neftyanye Kamni region, and 148 mg kg^{-1} in the Bakinskaya inlet.

Domestic sewage and industrial waste discharges from the large urban population around Baku have added to the stress on the coastal marine environment caused by oil pollution. Phytoplankton diversity in most of the Caspian is about 70 species, but in the western Caspian it has fallen from 74 to 40 species, and biomass from 8.7 to 2.1 g m^{-2}. The biomass of benthic organisms in coastal areas fell from 1724 to 21 g m^{-2} between 1961 and 1969.

In 1972, a series of measures was announced to provide sewage treatment for towns on the Volga, to treat industrial wastes, and to reduce oil pollution in the Azerbaijan region. These measures have not proved effective, however, and waste discharges and contamination levels in the southern Caspian remain high.

Experience since the 1950s has demonstrated the difficulty of carrying on a valuable fishery together with rapid industrialization, massive irrigation schemes, and exploitation of offshore oilfields in a sea subject to the natural pressures of high evaporation and a variable input of freshwater. The development of new production oilfields in the central and northern Caspian clearly provides a risk that this damage will be extended.

FURTHER READING

There is an enormous scientific literature, of variable quality, on matters relating to pollution of the sea. The following are some general publications relating to each chapter, and references to the sources of the text-figures and tables. The latter will give an entry to much of the relevant literature on matters discussed in this book.

Chapter 1

Earll, R.C. (1992). Commonsense and the precautionary principle; an environmentalist's perspective. *Mar. Pollut. Bull.*, **24**, 182–6.

Peterman, R.M. and M'Gonigle, M.M. (1992). Statistical power analysis and the precautionary principle. *Mar. Pollut. Bull.*, **24**, 231–4.

Royal Commission on Environmental Pollution, (1987). The best practicable environmental option. *Twelfth Report*, (Cm 310). HMSO, London.

Stebbing, A.R.D. (1992). Environmental capacity and the precautionary principle. *Mar. Pollut. Bull.*, **24**, 287–95.

Chapter 2

Gaugh, H.G. (1982). *Multivariate analysis in community ecology*. Cambridge University Press, Cambridge.

Peakall, D. (1992). *Animal biomarkers as pollution indicators*, Ecotoxicology Series, No. 1. Chapman & Hall, London.

Warwick, R.M. and Clarke, K.R. (1991). A comparison of some methods of analysing changes in benthic community structure. *J. mar. biol. Ass. U.K.*, **71**, 225–44.

Chapter 3

Champ, M.A. and Park, P.K. (ed.) (1989). *Monitoring waste management: science and policy*, (*Oceanic processes in marine pollution*, 3). Kreiger, Malabar FL.

Pearson, T.H. and Rosenberg, R. (1978). Macrobenthic succession in relation to organic enrichment and pollution of the marine environment. *Oceanogr. Mar Biol., Ann Rev.*, **16**, 229–311.

Proceedings of the International Symposium on Nutrient Dynamics in Coastal and Estuarine Environments (1995). *Ophelia*, **41**, **42**.

Chapter 4

Clark, R.B. (ed.) (1982). *The long-term effects of oil pollution on marine populations, communities and ecosystems*. Royal Society, London. [Also in *Phil. Trans. R. Soc. Lond.*, **B297**, 183–443].

National Research Council (1985). *Oil in the sea: inputs, fates and effects*. National Academy Press, Washington, D.C.

Royal Commission on Environmental Pollution (1981). Eighth Rept: *Oil pollution of the sea*, (Cmnd 8358). HMSO, London.

Wells, P.G., Butler, J.N., and Hughes, J.S. (ed.) (1995). *Exxon Valdez oil spill: fate and effects in Alaskan waters*. American Society for Testing and Materials, Philadelphia.

Chapter 5

Bryan, G.W. and Langston, W.J. (1992). Bioavailability, accumulation and effect of heavy metals in sediments with special reference to U.K. estuaries: a review. *Environ. Pollut.*, **76**, 89–131.

Furness, R.W. and Rainbow, P.S. (ed.) (1990). *Heavy metals in the marine environment*. CRC Press, Boca Raton, FL.

Mee, L.D. and Fowler, S.D. (ed.) (1991). Special issue on organotin. *Mar. Environ. Res.*, **32**.

Monteiro, L.R. and Furness, R.W. (1995). Seabirds as monitors of mercury in the marine environment. *Wat. Air, Soil Pollut.*, **80**, 851–70.

Chapter 6

Klamer, J., Loane, R.W.P.M., and Marguenie, J.M. (1991). Sources and fate of PCBs in the North Sea. *Water Sci. Technol.*, **24**, 77–85.

Newman, P.J. and Agg, A.R. (ed.) (1988). *Environmental protection of the North Sea.* Heinemann, Oxford (pp. 25–141 discuss organic chemicals).

Chapter 7

Bridges, O. and Bridges, J.W. (1995). Radioactive waste problems in Russia. *J. Radiol. Prot.*, **15**, 223–34.

Park, P.K., Kester, D.R., Duedall, I.W. and Ketchum, B.H. (1983). *Radioactive wastes and the ocean,* (*Wastes in the ocean,* vol.3). Wiley Interscience, New York.

Pentreath, R.J. (1980). *Nuclear power, man and the environment.* Taylor & Francis, London.

Chapter 8

ICES (1992). Report of the ICES Working Group on the effects of extraction of marine sediments on fisheries. *ICES Cooperative Research Report,* No.182. ICES, Copenhagen.

Ellis, D and Ellis, K. (1994). Very deep STD. *Mar. Pollut. Bull.*, **28**, 472–6.

Lucas, Z. (1992). Monitoring persistent litter in the marine environment on Sable Island, Nova Scotia. *Mar. Pollut. Bull.*, **24**, 192–9.

Richards, A.H. (1994). Problems of drift-net fisheries in the south Pacific. *Mar. Pollut. Bull.*, **29**, 106–11.

Robarts, M.D., Piatt, J.F., and Wohl, K.D. (1995). Increasing frequency of plastic particles ingested by seabirds in the subarctic north Pacific. *Mar. Pollut. Bull.*, **30**, 151–7.

Chapter 9

HELCOM (1993). First assessment of the state of the coastal waters of the Baltic Sea. *Balt. Sea Environ. Proc.*, No. 54. Helsinki Commission, Helsinki.

Jeftic, L., Bernhard, M., Demetropoulous, A., Fernex, F., Gabrielides, G.P., Gasparovic, F., Halim, Y., Orhon, D. and Saliba, L.J. (1990). State of the marine environment in the Mediterranean region. *UNEP Regional Seas Reports and Studies,* No.132. United Nations Environment Programme, Nairobi, Kenya.

North Sea Task Force (1993). *North Sea quality status report 1993.* Oslo and Paris Commissions, London.

SOURCES OF TEXT-FIGURES AND TABLES

Publishers and authors are warmly thanked for giving their permission for the use of material taken from their publications as follows.

Fig. 1.1 Royal Commission on Environmental Pollution (1984). *Tenth Report,* HMSO, London; **Fig. 2.1** Brown, B.E. (1972). *Effects of heavy metals on the physiology of selected invertebrates.* Unpublished thesis, University of London; **Fig. 2.2** Bryan, G.W. (1976). In *Marine pollution* (ed. R. Johnston), pp. 185–302. Academic Press, London; **Fig. 2.3** Bryan, G.W. (1976). In *Marine pollution* (ed. R. Johnston), pp. 185–302. Academic Press, London; **Fig. 2.4** Coombs, T.L. and George, S.G. (1977). In *Physiology and behaviour of marine organisms* (ed. D.S. McCluskey and A.J. Berry), pp. 179–87. Pergamon, Oxford; **Fig. 2.5** Cognetti, G. and Cognetti, G. (1992). *Inquinamenti e protezione del mare.* Calderini, Bologna; **Fig. 2.6 (a)** Anderson, D.S. (1973). *Embryology and phylogeny in annelids and arthropods.* Pergamon, Oxford; **(b)** Reish, D.J. *et al.* (1974). *Mar. Pollut. Bull.*, **5**, 125–6; **Fig. 2.7** Widdows, J. (1985). *Mar. Pollut. Bull.*, **16**, 129–34; **Fig. 2.8** Clarke, K.R. and Warwick, R.M. (1994). *Change in marine communities: an approach to statistical analysis and interpretation.* Natural Environment Research Council, Swindon; **Fig. 2.9** Warwick, R.M. (1986). *Mar. Biol.*, **92**, 557–62; **Fig. 2.10** Usero, J. *et al.* (1996). *Mar. Pollut. Bull.*, **32**, 305–10; **Fig. 2.11** Warwick, R.M. (1993). *Austral. J. Ecol.*, **18**, 63–80; **Fig. 2.12** Clarke, K.R. and Warwick, R.M. (1994). *Change in marine communities: an approach to statistical analysis and interpretation.* Natural Environment Research Council, Swindon; **Fig. 3.3** Royal Commission on Environmental Pollution (1972). *Third Report.* HMSO, London; **Fig. 3.4** Royal Commission on Environmental Pollution (1972). *Third Report.* London, HMSO; **Fig. 3.6** Wood, J.B. (1980). *Public Engineer*, **8**, 112–20; **Fig. 3.8** Shelton, R.G.J. (1971). *Mar. Pollut. Bull.*, **2**, 24–7; **Fig. 3.9** House of Lords Select Committee on the European Communities, Session 1985–6 (1986). *17th Report (HL219).* HMSO, London; **Fig. 3.10 (a)** and **(b)** Carmody, D.J. *et al.* (1973). *Mar. Pollut. Bull.*, **4**, 132–5; **(c)** White, H.H. *et al.* (1993). *Mar. Pollut. Bull.*, **26**, 49–51; **Fig. 3.11** Gray, J.S. (1981). *The ecology of marine sediments.* Cambridge University Press, Cambridge; **Fig. 3.12** Tatara, K. (1991). *Mar. Pollut. Bull.*, **23**, 315–19; **Fig. 3.13**

Riegman, R. *et al.* (1992). *Mar. Biol.*, **112**, 479–84; **Fig. 3.14** Steimle, F.W. and Sindermann, C.J. (1978). *Mar. Fish. Rev.*, **40**, 17–26; **Fig. 3.15** North Sea Task Force (1993). *North Sea quality status report 1993.* Oslo and Paris Commissions, London; **Fig. 3.16** Justić, D. (1987). *Mar. Pollut. Bull.*, **18**, 281–4; **Fig. 4.1** Shell (1994). Oil trading, *Shell Briefing Service*, No.2. Shell Petroleum, London; **Fig. 4.2** Wardley-Smith, J. (1976). *The control of oil pollution on the sea and inland waters.* Graham and Trotman, London; **Fig. 4.3** Sanders, P.F. and Tibbetts, P.J.C. (1987). *Phil. Trans. R. Soc. Lond.*, **B316**, 567–85; **Fig. 4.5** Whittle, K.J. (1982). *Phil. Trans. R. Soc. Lond.*, **B297**, 193–218; **Fig. 4.6** Wardley-Smith, J. (1976). *The control of oil pollution on the sea and inland waters.* Graham and Trotman, London; **Fig. 4.8** Wardley-Smith, J. (1976). *The control of oil pollution on the sea and inland waters.* Graham and Trotman, London; **Fig. 4.9** Wardley-Smith, J. (1976). *The control of oil pollution on the sea and inland waters.* Graham and Trotman, London; **Fig. 4.10** North, W.J. *et al.* (1964). *Symp. pollut. mar. micro-org. prod. pétrol.*, pp. 335–54, CIESMM, Monaco, and Nelson-Smith, A. (1972). *Oil pollution and marine ecology.* Elek, London; **Fig. 4.11** Bourne, W.R.P. (1976). In *Marine pollution* (ed. R. Johnston), pp. 403–502. Academic Press, London; **Fig. 4.12** Kingston, P.F. (1987). *Phil. Trans. R. Soc. Lond.*, **B316**, 545–65; **Fig. 4.13**, Kingston, P.F. (1987). *Phil. Trans. R. Soc. Lond.*, **B316**, 545–65; **Fig. 5.1** Cunningham, F.A. and Tripp, M.R. (1975). *Mar. Biol.*, **31**, 311–9; **Fig. 5.2** Häkkinen, I. and Häsänen, E. (1980). *Ann. Zool. Fenn.*, **17**, 131–9; **Fig. 5.3** Bryan, G.W. and Gibbs, P.E. (1983). *Occ. Pap. mar. biol. Ass. U.K.*, No.2. Marine Biological Association of the U.K., Plymouth; **Fig. 5.4** Bryan, G.W. (1976). In *Marine pollution* (ed. R. Johnston) pp. 185–302. Academic Press, London; **Fig. 5.5** Murozumi, M. *et al.* (1969). *Geochim. Cosmochim. Acta*, **33**, 1247–94; **Fig. 5.6** Lee, J.A. and Tallis, J.H. (1973). *Nature, London.*, **245**, 216–8; **Fig. 5.7** Gerlach, S.A. (1981). *Marine pollution: diagnosis and therapy.* Springer, Heidelberg; **Fig. 5.8** Newell, R.C. *et al.* (1991). *Mar. Pollut. Bull.*, **22**, 112–8; **Fig. 6.2** Smokler, P.E. *et al.* (1979). *Mar. Pollut. Bull.*, **10**, 331–4; **Fig. 6.3** Halcrow, W. *et al.* (1974). *Mar. Pollut. Bull.*, **5**, 134–6; **Fig. 6.4** Goerke, H. *et al.* (1979). *Mar. Pollut. Bull.*, **10**, 127–33; **Fig. 6.5** Ernst, W. and Goerke, H. (1974). *Mar. Biol.*, **24**, 287–304; **Fig. 6.6** Mellanby, K. (1967). *Pesticides and pollution.* Collins, London; **Fig. 6.7** Patin, S.L. (1982). *Pollution and the biological resources of the oceans.* Butterworth, London; **Fig. 6.8** Ernst, W. (1980). *Helgol. Meeresunters.*, **33**, 301–12; **Fig. 6.9** data from Butler, P.A. (1972). In *Marine pollution and sea life* (ed. M. Ruivo), pp. 262–6. Fishing News (Books) Ltd, London; **Fig. 6.10** North Sea Task Force (1993). *North Sea quality status report 1993.* Oslo and Paris Commissions, London; **Fig. 6.11** MAFF (1994). *Aquatic Environment Monitoring Report*, No.40. MAFF Directorate of Fisheries Research, Lowestoft; **Fig. 7.2** Hodge, V.F. *et al.* (1973). *Proc. symp. radioactive contamination of the marine environment*, pp. 263–76, IAEA, Vienna; **Fig. 7.3** Føyn, L. (1994). *ICES Comm. Meeting, CM1994/E.16.* ICES, Copenhagen; **Fig. 7.4** Radioactive Waste Management Committee (1983). *Fourth Annual Report*, HMSO, London; **Fig. 7.5** MAFF (1994). *Aquatic Environment Monitoring Report*, No.42. MAFF Directorate of Fisheries Research, Lowestoft; **Fig. 7.6** Føyn, L. (1994). *ICES Comm. Meeting, CM1994/E.16.* ICES, Copenhagen, and Bridges, O. and Bridges, J.W. (1995). *J. radiol. Prot.*, **15**, 223–34; **Fig. 7.7** Fowler, S.W. and Gaury, J.C. (1971). *Nature, Lond.*, **266**, 827, and Pentreath, R.J. (1980). *Nuclear power: man and the environment.* Taylor and Francis, London; **Fig. 7.8** Woodhead, D.S. (1980). *Helgol. Meeresunters.*, **33**, 122–37; **Fig. 7.9** data from Webb, G.A. *et al.* (1991). National Radiological Protection Board, *Rept. NRPB-M286.* NRPB, Didcot; **Fig. 8.1** ICES (1992). *ICES Cooperative Research Report*, No.182. ICES, Copenhagen; **Fig. 8.2** Rosenberg, R. (1977). *Mar. Pollut. Bull.*, **8**, 102–4; **Fig. 8.3** Pointer, I.R. and Kennedy, R. (1984). *Mar. Biol.*, **78**, 335–52; **Fig. 8.4** Anon. (1970). *Mar. Pollut. Bull.*, **1**, 115: **Fig. 8.5 (a)** Portmann, J.E. (1970). *J. mar. biol. Ass. U.K.*, **50**, 577–91; **(b)**, **(c)** Howell, B.R. and Shelton, R.G.J. (1970). *J. mar. biol. Ass. U.K.*, **50**, 593–607; **Fig. 8.6** Dixon, T.R. and Dixon, T.J. (1979). *Mar. Pollut. Bull.*, **10**, 352–7; **Fig. 8.7** Thorhaug, A. (1978). *Mar. Pollut. Bull.*, **9**, 191–7; **Fig. 8.8** Langford, T.E. (1983). *Electricity generation and the ecology of natural waters.* University of Liverpool Press, Liverpool; **Fig. 9.1** Korringa, P. (1968). *Helgol. wiss. Meeresunters.*, **17**, 127–40; **Fig. 9.2** Aksnes, D.L. *et al.* (1995). *Ophelia*, **41**, 5–36; **Fig. 9.3** North Sea Task Force (1993). *North Sea quality status report*, 1993. Oslo and Paris Commissions, London; **Fig. 9.4** North Sea Task Force (1993). *North Sea quality status report*, 1993. Oslo and Paris Commissions, London; **Fig. 9.5** Anon, (1980). *Ambio*, **9**, 114–5; **Fig. 9.6** Fonselius, S. (1981). *Mar. Pollut. Bull.*, **12**, 187–94; **Fig. 9.7** Andersin, A.B. (1978). *Kieler Meeresforsch.*, suppl. **4**, 232–52; **Fig. 9.8** HELCOM

(1993). *Balt. Sea Environ. Proc.*, No.54. Helsinki Commission, Helsinki; **Fig. 9.9** HELCOM (1993). *Balt. Sea. Environ. Proc.*, No.54. Helsinki Commission, Helsinki; **Fig. 9.10** Heimer, R. (1997). *Ambio*, **6**, 313–6; **Fig. 9.11** Le Lourd, P. (1977). *Ambio*, **6**, 317–9; **Fig. 9.12** Aubert, M. (1994). *La Méditerranée: la mer et les hommes.* Editions de l'environnement, Paris; **Fig. 9.13** Thorhaug, A. (1981). *Ambio*, **10**, 295–8; **Fig. 9.14** Hughes, T.P. (1994). *Science*, **265**, 1547–51; **Fig. 9.15** Rogers, C.S. (1985). *Proc. 5. internat. coral reef congr.*, **6**, 491–6; **Fig. 9.16** adapted from Zenkevitch, L.A. (1957). In *Treatise on marine ecology and palaeontology*, **1**, 891–916. Geol. Soc. Am., New York; **Fig. 9.17** Karpinsky, M.G. (1992). *Mar. Pollut. Bull.*, **24**, 384–9; **Fig. 9.18** Karpinski, M.G. (1992). *Mar. Pollut. Bull.*, **24**, 384–9.

Table 2.1 Gray, J.S. and Ventilla, R.J. (1973). *Ambio*, **2**, 118–21; **Table 2.2** Bryan, G.W. *et al.* (1985). *Mar. biol. Assoc. U.K., Occ. Publ.*, No.4. Marine Biological Association of the U.K., Plymouth; **Table 2.3** Bonney, A.D. *et al.* (1959). *Biochem. Pharmacol.*, **2**, 37–49; **Table 2.4** White, H.H. and Champ, M.A. (1983). In *Symp. hazardous and industrial solid waste testing* (ed. R.A. Conway and W.P. Gulledge), pp. 200–312. Amer. Soc. Testing and Materials, Philadelphia; **Table 4.1** Steering committee for the petroleum in the marine environment update (1985). *Oil in the sea: inputs, fates and effects.* National Academy Press, Washington, D.C., and Ambrose, P. (1991). *Mar. Pollut. Bull.*, **22**, 262; **Table 4.2** Exxon (1989). *Review of oil spill occurrences and impacts.* Exxon Production Research Co., Houston TX, and ITOPF (1995). *Oil spill database.* International Tanker Owners Pollution Federation Ltd, London, with later additions; **Table 4.3** B.P. (1977). *Our industry petroleum.* British Petroleum, London; **Table 4.4** Royal Commission on Environmental Pollution (1981). *Eighth Report.* HMSO, London; **Table 4.5** Royal Commission on Environmental Pollution (1981). *Eighth Report.* HMSO, London; **Table 5.1**

Pacyna, J.M. (1986). In *Toxic metals in the atmosphere* (ed. J.O. Nriagu and C.I. Davidson), pp. 33–52. Wiley Interscience, New York; **Table 5.2** Buat-Ménard, P. (1986). In *The role of air–sea interchange in geochemical cycling* (ed. P. Ménard), pp. 477–96. Reidel, Dordrecht; **Table 5.3** North Sea Task Force (1993). *North Sea quality status report 1993.* Oslo and Paris Commissions, London; **Table 5.4** Keckes, S. and Miettinen, J.K. (1972). In *Marine pollution and sea life* (ed. M. Ruivo), pp. 276–89. Fishing News (Books) Ltd, London, Craig, P.J. (1986). *Organometallic compounds in the environment: principles and reactions.* Longman, London, and DeVito, S.C., (1995). In *Encyclopedia of chemical technology*, 4th edit. (ed. M. Howe-Grant), **16**, 212–28. Wiley Interscience, New York; **Table 6.1** GESAMP (1989). *GESAMP Reports and Studies*, No.38, and Preston, M.R. (1992). *Mar. Pollut. Bull.*, **24**, 477–83; **Table 6.2** GESAMP (1989). *GESAMP Reports and Studies*, No. 38. International Maritime Organization, London, and Preston, M.R. (1992). *Mar. Pollut. Bull.*, **24**, 477–83; **Table 6.3** Smokler, P.E. *et al.* (1979). *Mar. Pollut. Bull.*, **10**, 331–4; **Table 7.1** International Atomic Energy Authority (1976). *Technical Reports Series*, No.172, IAEA, Vienna, units adjusted; **Table 7.2** Woodhead, D.S. (1976). *Proc. symp. radioactive contamination in the marine environment*, pp. 455–515. IAEA, Vienna, units adjusted; **Table 7.3** *Report of the independent review of disposal of radioactive waste in the northeast Atlantic* (1984). HMSO, London (1984), units adjusted; **Table 7.4** International Atomic Energy Authority (1976). *Technical Report Series*, No.172. IAEA, Vienna, units adjusted; **Table 7.5** Woodhead, D.S. (1984). In *Marine ecology* (ed. O. Kinne), **5**, 1112–287. Wiley, London, units adjusted; **Table 7.6** Camplin, W.C. and Aarkrog, A. (1989). *Data Report*, No.20. MAFF Directorate of Fisheries Research, Lowestoft; **Table 8.1** North Sea Task Force (1993). *North Sea quality status report 1993*, Oslo and Paris Commissions, London; **Table 8.2** Dixon, T.R. and Dixon, T.J. (1979). *Mar. Pollut. Bull.*, **10**, 352–7.

INDEX